The Mysterious and the Occult from Newton to the Victorians

Michael Punter

The Mysterious and the Occult from Newton to the Victorians

Strange Powers

*Do not let us be diverted from facts by the incompleteness
or presumptuousness of our theories.
Oliver Lodge, British physicist, 1924.*

*For my parents, Roy and Valerie Punter, with love.
Thanks to Lisa, Will and Manny, Tom Goodman Hill, Amanda
Grace, Professor Ben Poore, Dr Mike Woolf, the staff and students
at CEA CAPA London.*

Contents

1. Introduction — 1
2. Prologue: Newton and the Mysterious — 7
3. Forces in the Body — 17
4. Forces in the Mind — 31
5. The Electric Miracle — 47
6. Sensitives, Mediums and Conduits — 65
7. The Brief Life of Psychic Force — 77
8. A Psychic Body — 89
9. An Eternal Mind — 105
10. Ether/Or… — 117

| 11 | Epilogue: Even Stranger Things | 127 |

Bibliography 133

Author Index 137

Subject Index 141

CHAPTER 1

Introduction

Abstract In the introduction, I describe my academic interests and explain the reasons for writing the book.

Keywords Magic · Witchcraft · Theatre · Literature · Ether · God Particle

For the last five years, I've had the pleasure of teaching an undergraduate class titled 'Witchcraft and Magical Performance' for an American education company. My classes are held in London and the students come from across the United States. Their range of majors is diverse and includes Psychology, Economics, Business, English Literature and Drama. I begin the course by getting everyone in the class to share a mysterious experience they've had. Every student has one, be it an alarming coincidence, ghostly encounter, déjà vu or just an eerie feeling about a place. In my experience, most people have had some kind of encounter with something they struggle to explain. Mystery is always with us.

When teaching, I always try to find current analogues for historical phenomena. For example, I teach that social media is a place of magic in that it's full of tricks and scams involving love or money. Online, people can transform themselves into others to distract a victim, or to extract payment from them. Although a seventeenth-century witch would be shocked by our technology (and undoubtedly consider it magical) the motives and sense of craft in an online scam and an Early Modern spell are essentially the same. I feel that, once I've established a modern equivalent, it makes my class more invested in the historical journey. In the final class, I talk about the Large Hadron Collider at CERN, The European Organization for Nuclear Research, a giant particle accelerator, the biggest and greatest machine ever made by humans, located beneath the border between France and Switzerland. It was designed, in part, to identify the strangest and most exotic particles in the universe. From there, I trace a line back to 1887 and the greatest failed experiment of the nineteenth century: Michelson and Morley's attempt to prove the existence of a mysterious substance in space called 'the ether'. This ether was, they surmised, a mysterious fluid that comprised most of the universe and was the conduit for the light wave to travel from the sun to earth. But their experiments failed to detect it at all and, in the summer of 1887, a painstakingly constructed model of the universe, inspired by Newton's theory of gravity proposed two centuries previously, began to unravel. It was eventually superseded by Einstein's special relativity which led, in time, to the greatest mystery of all: the world of quantum mechanics. If the Collider hadn't detected the existence of the exotic particle known as the Higgs Boson or 'God Particle' in 2012, then it would probably have been as disconcerting for modern physicists as the elusiveness of the ether was to their Victorian counterparts. The mystery of the cosmos would have been deepened to the point of seeming unfathomable.

I'm not a scientist. My academic interests have been in literature and drama for over two decades, and my PhD thesis was on the work of a Victorian actor called Henry Irving. But my interest in the subject of mysterious forces began there, with Irving frequently described as possessing a strange, almost telepathic, control over his audience. In his greatest stage success, *The Bells*, produced in 1871, Irving played the part of a French burgomaster who was 'mesmerized' against his will and forced to re-enact a murder he'd committed 15 years previously.[1] The performance was a sensation and addressed a range of anxieties, conscious and unconscious, about how much control Victorians actually had over the forces at work in their minds and bodies. Irving's personal manager was the author Bram Stoker. When they first met in Dublin in 1876, Stoker truly believed that Irving had made him—a hard bitten and world-weary journalist—subject

---

[1] This is explored more fully in Chap. 4.

to some mysterious power that had acted upon him and reduced him to a sobbing 'hysteric'. Twenty-one years later, psychic control was one of the powers of Stoker's greatest creation, Count Dracula. But not every observer believed in the invisible power of Irving. The contemporary critic William Archer viewed Irving as an absurd practitioner of melodrama whose supposed power was simply a handful of hackneyed, second-hand gestures that belonged in another era. His revered charisma was like a magic trick that Archer knew the secret of. Yet, by the end of the nineteenth century, a range of new terminology was being used to describe performers: 'magnetic', 'electrifying' and—in Irving's case—'mesmerizing'. Whether or not an individual experienced such forces in this period depended very much on their point of view and social position.

So, what were the mysterious forces that were discovered (or *thought* to have been discovered) in the two centuries after Newton? I'm particularly interested in what people *imagined* them to be before their identity was established or before they were debunked entirely. I'm also interested in the courageous, clever and sometimes foolhardy scientists who risked their social status and professional standing by investigating them and proposing imaginative hypotheses. Chapters 3 and 4 describe the 'discoveries' of Franz Anton Mesmer and the tortuous journey towards what we now call 'hypnotism', and I take some time to examine the points of interaction between philosophy and science in the aftermath of Mesmer, and his impact on popular culture. In Chap. 5, I consider electricity, and the nature of the process Mary Shelley thought was happening in the animation of her famous creature in *Frankenstein*. Far from being a convenient lightning strike, it's a vague concoction of alchemical processes that might be termed 'occult', a word whose meaning I will explain and contextualize shortly. The Italian scientist Luigi Galvani believed that he'd discovered a divine spark in the twitching legs of a dead frog, but his rival Alessandro Volta concluded it was simply the dispassionate passing through of electrical current. By not clarifying the nature of the force at work, Shelley made it possible for her audience to engage and imagine with the greatest philosophical questions. Chapters 6, 7 and 8 explore the influence of new ideas about nature on a generation of progressive scientists who were from outside of the privileged, Oxbridge elites. One of these, the chemist William Crookes, believed there was such a thing as 'psychic force' and that the mind could not only interact with the material world, but that it could even survive physical death. The largely unsung British physicist Oliver Lodge never gave up on the existence of the ether of space, despite the failure of Michelson and Morley's experiment. An evaluation of his efforts to define the qualities of the ether, and to prove its interactive nature, make up Chaps. 9 and 10. These investigators were astonishing thinkers, and, during the writing of this book, I found myself constantly asking if they simply saw something we cannot. Were their imaginations broader, unconstrained by the boundaries modern materialism imposes upon us? Whatever the truth of it, I think it's likely that we will always seek out mysterious forces and powers because we need to 'solve' them, because the possibility of an unintelligible cosmos is too much to bear. Or perhaps it's simply further evidence of our endless capacity to *imagine*.

We might give strange forces a more human identity, and maybe it's because they *are* deeply connected to us. Since I first read the book as a teenager, I have always loved this quote from the end of Primeval Night, Part One of Arthur C. Clarke's *2001: A Space Odyssey*: 'Presently he (humankind) invented philosophy, and religion. And he populated the sky not altogether inaccurately with gods'.[2] The earliest attempts to understand the forces at work in nature involved giving them the faces of divinities. It was an intuitive and credible solution for ancient humans. I conclude this book with an (brief) attempt to understand an area that's far harder to make familiar: quantum physics. This is *our* mystery, the one we seem blessed and cursed to try to unravel. Perhaps it should be left alone, but I feel it would be impolite not to at least *try* to engage with the many, multifaceted particles that are streaming through me as I write this.

This journey might be a slightly bumpy one. My students say I have an annoying habit of going off at a tangent. This is true, but it's only because I suspect that all things might be secretly connected. Perhaps it's better to suggest that everything may be 'entangled'? I studied the humanities, but I love the sciences and especially value anything that brings them together. So please be prepared for wanderings into the worlds of literature, philosophy and history, and to encounter heartfelt truth seekers and downright frauds. To borrow the title of a successful film, we experience 'everything, everywhere, all at once', just as our ancestors did. And like them, we should allow ourselves to be provoked, surprised and overwhelmed by the richness of the world around us.

I'd like to thank the scholars whose work has inspired me, particularly Janet Oppenheim, Molly McGarry, Iwan Rhys Morus and Richard Noakes.

---

[2] Arthur C. Clarke, *2001: A Space Odyssey: 50th Anniversary Edition* (London: Orbit, 2018) page 31.

They have all written with great insight about the eighteenth and nineteenth centuries and emphasized the many areas of overlap between 'legitimate' science and 'occult' practice. They've also written of the agency that emergent forces, or *ideas* of forces, granted to marginalized social groups. If I have managed to glimpse anything new in this field, it's only because such fine scholars have illuminated the path.

CHAPTER 2

Prologue: Newton and the Mysterious

Abstract Here, I outline my approach to the exploration of mysterious and occult forces. Terms are defined and the interpretative framework established. A brief history of such forces is given, and their manifestations are contextualized. The consequences of Newton's theory of gravitation are described and a list of the forces to be examined is presented.

Keywords Mysterious • Occult • Force • Gravity • Isaac Newton • Science • Magic • Alchemy

> But hitherto, I have not been able to discover the cause of these properties of gravity from phenomena, and I do not frame hypotheses (Isaac Newton, *Principia Mathematica*).[1]

In 1687, Isaac Newton published the *Principia Mathematica*. In this work, he explained his Three Laws of Motion before going on to describe the operation of gravitational force. According to his first biographer, William Stukeley, Newton's newly discovered power 'extends itself thro'

---

[1] Isaac Newton, trans. A. Motte, *Newton's Principia: The Mathematical Principles of Philosophy* (New York: Daniel Ader, 1846) page 506.

© The Author(s), under exclusive license to Springer Nature Switzerland AG 2024
M. Punter, *The Mysterious and the Occult from Newton to the Victorians*, https://doi.org/10.1007/978-3-031-67882-0_2

the universe'.² It explained the motions of heavenly bodies and 'kept the planets from falling upon one another or dropping together into one centre'.³ It also described the role of the sun's gravity in the shaping of the earth and the effects of the moon upon the tides. This discovery challenged the conceptual separation between 'celestial' and 'terrestrial' mechanics that had existed theoretically since ancient times. Newton's theory brought heaven and earth into unity.

Newton's discoveries, along with those of Hooke, Boyle and Halley, created a framework for scientific method and investigation. However, Newton was far from being a conventional figure. In faith, he believed in the Arian doctrine, a Christian heresy that denied the Trinity and relegated Christ to the position of first created being of God the Father. Although Charles II's England was more tolerant than many other European countries, Newton had to downplay his beliefs and required special dispensation from the monarch to advance his career. Additionally, a substantial amount of Newton's research concerned alchemy, the ancient and mysterious art that sought—among other arcane objectives—to find the mechanism for the transmutation of base metals into gold. Newton's search for forces that underpinned and bound the physical world drew strongly on a tradition that could be termed 'occult'. As Mark A. Waddell has written in *Magic, Science and Religion in Early Modern Europe*:

> Newton, then, began by considering the possibility that matter was imbued with some kind of active principle or attractive virtue that could explain why particular phenomena occur in nature.⁴

Newton's faith in God was profound, and it's possible to see his research in both gravity and light as an attempt to introduce a spiritual property into a material world operating upon strictly mechanical principles. As Waddell states:

> If one wanted to simplify Newton's great achievement in physics and cosmology, they might say that he took the mechanical, materialist philosophies

---

[2] William Stukeley, *Memoirs of Sir Isaac Newton's Life* (London: Royal Society, 1752) page 14.

[3] Ibid.

[4] Mark A. Wadell, *Magic Science and Religion in Early Modern Europe* (Cambridge: CUP, 2021) page 185.

of the seventeenth century and added to them the concept of an immaterial and invisible force.⁵

It's necessary to pause here and give some sense of the development of the ideas I'll be discussing. Newton regarded gravitational force as the guiding hand of his singular God, but there had been explanations for the movement of objects, vast and tiny, across millennia, beginning with the actions of multiple gods. The Babylonians made complex computations of the positions of heavenly bodies to predict terrestrial events, linking the world 'above' with the world 'below' perhaps for the first time. This is evidenced by calculations of the position of the planet Venus on clay tablets from the second millennium BCE. In China in the fifth century BCE, officials were using astronomical data to ascertain the times of the seasons. In western Asia, on the coast of what's now Turkey, Thales of Miletus attempted to construct theories regarding the natural world that were based upon observable laws rather than mythological explanations. His peer, and probable student in the sixth century BCE, was Anaximander. Anaximander's model of the cosmos placed earth at its centre but, remarkably, suggested that the heavenly bodies were being hauled by vast wheels filled with fire. These respective, concentric wheels contained apertures through which the sun and moon could be viewed. Even more remarkably, Anaximander taught that the earth was floating free in space. It's hard to overstate the importance of this groundbreaking idea. As Carlo Rovelli has written:

> understanding that the Earth is a stone that floats unsupported in space, with the same heaven underneath it as the one we see above – this is a huge step forward conceptually. And this is Anaximander's contribution.⁶

Mythologies across the world had proposed that the earth was supported by the shoulders of a hero or a huge, divine entity, but Anaximander's concept broke decisively with such explanations. Equally striking was his idea of the 'apeiron'. Rovelli describes this as 'the indefinite or infinite' from which all of nature derives:

---

⁵ Ibid., page 186.
⁶ Carlo Rovelli, *Anaximander and the Nature of Science* (London: Allen Lane, 2023) page 49.

> The world came into being when hot and cold separated from the apeiron. This separated the cosmic order. A ball of flame grew around the air and the Earth…The ball then broke apart and was confined inside the wheels that form the Sun, the Moon, and the Stars.[7]

Two centuries later, Aristotle similarly employed the circle as the mode of celestial operation. The heavens were in motion but the earth was still, and the movement of the celestial entities was attributed to the operations of 'prime movers'. In terrestrial mechanics, objects moved because both the moved and the mover had the 'potential' for such action.[8] Earthly substances were comprised of the four elements: earth, air, fire and water and the movement of these elements was attributed to their need to return to their 'natural resting places',[9] so stones fall to earth, and flames reach up to heaven. There was an additional, fifth element in which the celestial bodies moved, and this was known as the 'ether'.[10]

Aristotle's *Physics* was based on observation and intuition and it's almost completely wrong. And yet the domination of his thought, yoked to Catholic theology in the European Middle Ages in a conceptual alliance that probably would have both shocked and amused him, meant that challenges to this world view were hard to sustain. But the challenges came in the Early Modern period in the form of Copernicus, Galileo and Kepler. As Newton conceded, he was 'standing on the shoulders of giants' in his own attempt to solve the mysteries of the universe.

The word 'force' has a range of meanings. The *Cambridge Dictionary* begins its definition with a strikingly simple description: '(a measure of) the influence that changes movement'.[11] I'll be using it to consider a variety of ways in which the world was perceived to be influenced by something hidden or mysterious from Newtonian gravity at the end of the seventeenth century to the idea of the ether of space at the end of the nineteenth century. I'm interested in the actions of forces upon physical bodies, but also in the idea of *influence* as a force upon the mind that causes it to act in a certain way, sometimes a way that was not considered to be 'natural' or 'normal'.

---

[7] Ibid., page 31.

[8] Aristotle, trans. R P Hardie and R K Gaye *The Physics. Writings on Natural Philosophy* with a New Introduction by Dr. James Lees (London: Flame Tree 2023) page 91.

[9] Ibid. page 83.

[10] This deserves, and gets, a chapter to itself.

[11] https://dictionary.cambridge.org/dictionary/english/force

During the two hundred years after Newton, many would seek out, and claim knowledge of, mysterious and invisible forces. If medieval alchemists had sought to harness hidden powers and reveal the secret matter at the heart of the universe, then their interests could hardly be said to have gone out of fashion in the eighteenth and nineteenth centuries. In the scholarship of the last few decades, and thanks particularly to the insights of the American historian Betty Jo Teeter Dobbs,[12] Newton has been considered as both 'First Scientist' and as 'Last Alchemist'. In truth, Newton was a philosopher whose many interests coalesced around two key questions: 'What is the nature of matter?' and 'What is the role or agency of God in the universe?' Far from being a separate aspect of his study, Newton's fascination with alchemy, his attempt to understand the forces at work in creation, was as significant as his interests in light and gravity, and a part of the same great project. In 1669, the year he became Lucasian Professor of Mathematics at Cambridge, he purchased two furnaces for experimenting with metals and a set of alchemical manuscripts. As Michael White has written of him:

> He must have come to the conclusion that, if there was an underlying principle to the occult, then it was this: that, if studied from an intellectual standpoint, the occult would act as a glue, a unifier of fundamental principles; that it could be rationalized and logicalized.[13]

Newton sought to maintain God within an ordered universe. For him, divine power passed from the creator, through the mystery of the first-created being, the Christ, and into the world of matter. This spiritual force was activating and energizing, granting motion and vibrancy to all things. During the subsequent decades, thinkers of the Enlightenment would challenge the role of God in the universe or at least, an idea of God *based on scripture*. But the concept of active and invisible forces would continue to inspire scientists, pseudo-scientists and philosophers into the eighteenth century and beyond. What interests me here are attempts to approach and resolve concepts of *mystery*—particularly during the nineteenth century—in order to increase the agency of those lacking opportunity and the resources for self-improvement. The idea of secret, unseen forces within

---

[12] As Waddell points out, Teeter Dobbs was the first scholar to demonstrate the importance of the problem of 'inert matter' in Newton's thought.
[13] Michael White, *Isaac Newton: The Last Sorcerer* (London: Fourth Estate, 1997) page 107.

the universe inspired many to study them, explore them and use them to advance their own scientific, spiritual and political projects.

The two hundred or so years that separate Newton's theory of gravity and the non-discovery of the ether of space represented a period of revelation in which art and science appeared, fleetingly, to be reconciled, and the project of uncovering universal and unifying principles seemed tantalizingly credible. This book is a journey across that time, and into a world that continues to fascinate me. The conceptual lens through which I've decided to view this remarkable range of phenomena is 'the mysterious', from the ancient Greek word 'mysterion' that originally meant a secret rite, derived from the term 'mystes', a word denoting initiation into a (usually magical) practice. It occurs many times in the Koine Greek of the New Testament, particularly in the *Letters* of Paul and in the *Acts of the Apostles*. Perhaps the most quoted use is in 1 *Corinthians*: 15: 50–52:

> Listen, I tell you a mystery: We will not all sleep, but we will all be changed in a flash, in the twinkling of an eye, at the last trumpet. For the trumpet will sound, the dead will be raised imperishable, and we will be changed.[14]

The term is useful for a study such as this because it engages with the idea of a process that leads to greater knowledge and revelation. For ancient Greeks, access to mystery probably entailed a sequence of trials or rites of passage in a devotional space that enabled the supplicant of a god to travel—physically and psychically—through realms of greater knowledge. The devotee gained deeper insight into the nature of the divine realm. To a significant degree, I will argue, this theological view of mystery continued to inform the investigations of the scientists of the eighteenth and nineteenth centuries. I believe 'mystery' can be deployed effectively in this study because almost all of the philosophers and scientists whose work I'm going to describe were practising Christians, although several were in profound dispute with the dominant iteration of the faith of their time. Not only were they largely non-conformist, but they were also, at least initially, from outside of the social elites of the period. Many were first-generation scholars, the sons of craftsmen who saw in the practice of their parents something remarkable or even magical. This encouraged them to investigate the mysteries of the world around them, inspired them to seek hidden

---

[14] New International Version, 2015. https://www.biblegateway.com/passage/?search=1%20Corinthians%2015&version

causes and served as the first rung on a ladder to professional and social advancement. Michael Faraday, one of the most remarkable minds of the nineteenth century, was the son of a blacksmith whose first experience of mystery was the heating and shaping of metal in his father's forge. He never lost his sense of awe when he regarded the processes of transformation.

The other term I will use is rather more freighted: 'the occult'. This word has attracted a number of associations, largely negative, over time. It's often associated with aspects of malign or 'black' magic and the kind of activity seen in horror films with a supernatural theme. When I refer to the 'occult' in this book, I am generally using the meaning derived from Latin: 'hidden'. This iteration covers a range of invisible and intangible phenomena, real and imagined, proved and debunked (and perhaps yet to be identified?).

From Newton onwards, a group of mostly men from outside of the educational and social elites set their minds to investigating an extensive range of mysteries that encompassed physics, chemistry, biology, psychology and more. They engaged in practical, experimental work partly based on a Newtonian example but also inspired by how they had seen the previous generation work—largely with their hands in professions that made material things. From Newton's theory of gravitation to the disproving of the existence of the ether of space, the investigation of the mysteries of the universe moved decisively away from religious authorities and into the hands of radical, practical young experimenters. The search for truth became, to some extent, democratized in the period of the Industrial Revolution and agency was given to a new class of investigators. Although their respective journeys were influenced by the political upheavals that punctuated the period, they often found themselves attempting to reconcile a scriptural concept of a benign deity with the strange new world they were uncovering experimentally. The result was an imaginative collision of ideas that we in the twenty-first century are still struggling to come to terms with. What's particularly exciting about the scientists of this time is the expansive nature of their thought in which material and metaphysical ideas were at war with each other and yet also seeking union and resolution. It was acceptable, although not always wise, to consider mysterious phenomena such as mesmerism, telepathy and telekinesis along with more legitimate and established concepts. Indeed, many of the scientists I will consider felt compelled, as a matter of scientific duty, to investigate phenomena that working people had encountered and found to be troubling.

It was only towards the end of the nineteenth century that the focus of investigation narrowed and decisively labelled certain fields of mystery as unacceptable or tainted. This, I will argue, disconnected many from the narratives of philosophy and science, partly closing those areas (especially in England) to all but a university-based elite. Such obstructions to knowledge are certainly still in place today. Indeed, they may be more formidable to the outsider than ever before.

Before we continue, I'll outline the structure of this book in a little more detail. It's quite straightforward and moves, generally, in chronological order with allowances for certain important connections between different fields over time. It begins with animal magnetism and the attempts to expand the areas of investigation revealed by Newton. How might vast, planetary forces affect humans and were there other powers that could be harnessed? I'll partly revise the reputation of Anton Mesmer, setting him alongside his eighteenth-century near-contemporary Emanuel Swedenborg—a man of science who became a mystic to the disappointment of the great philosopher Immanuel Kant—but whose works in both fields provide a fascinating example of the many areas of entanglement between faith and science. I'll proceed to examine the career of John Elliotson and the uses of the theories of Mesmer's students to create new psychological models along with the practical applications of trance-based anaesthesia. I'll then look at ideas of 'vitalism' with a particular focus on Karl von Reichenbach and the theory of an all-embracing 'Odic force' that all other forces were merely an expression of. As I've already mentioned, Mary Shelley's *Frankenstein* is the focus for a range of anxieties about the nature and practical uses of electricity, and I'll explore the use of it as both a mysterious power source and an equally mysterious means of communication. This discussion will also inform chapters on theories of telepathy and telekinesis, after the mid-nineteenth-century preoccupation with the ghostly and the possible survival of consciousness has been encountered. Our final topic will be the quest to discover the physical properties of space in the form of the deeply mysterious substance referred to by several ancient philosophers: the ether. This project, I will argue, represented a heroic attempt to join mind and matter to create a universal theory of everything that would inform the approach to mystery in the twentieth century and beyond. Without the challenge of the ether, Einstein's Special Theory of Relativity would probably not have been possible. As Carlo Rovelli has written in his work on Anaximander:

> The aspect of science that I seek to illuminate…is its critical and rebellious ability to reimagine the world again and again.[15]

Some of the ideas offered by scientists in this book were incorrect, and yet their imaginative attempts to explain the world were expansive and inspiring. This led many people from outside of the elites to openly question and investigate the phenomena around them. The consequences of those investigations were profound both for the scientists and for their increasingly global audience.

---

[15] Ibid. page xii.

CHAPTER 3

Forces in the Body

Abstract In this chapter, I consider some responses to Newton's theory of gravitation, with a particular focus on the work of Franz Anton Mesmer. If gravity permeates the universe, then what is its relationship to the human body? Mesmer's thought is placed in context with that of a number of other eighteenth- and nineteenth-century thinkers, including Swedenborg, Kant and Schopenhauer, and also set alongside ideas from Nature Philosophy.

Keywords Mesmerism • Nature • Magnetism • Influence • Will • Transcendental Idealism • Swedenborg • Kant • Schopenhauer

> There exists a reciprocal influence between the heavenly bodies, the earth, and all living beings. (Franz Anton Mesmer, *Aphorisms*)[1]

The discoveries of Newton were hugely influential across a range of fields and brought a critical weight to arguments for the importance of the

---

[1] Justinus Kerner, trans. A M H Watts and Preface by C Mingins. *Franz Anton Mesmer, the Discoverer of Animal Magnetism* (Hidden Tarn Editions, 2020) page 60. Kerner claims to have transcribed Mesmer's hand-written list of aphorisms into his biography. The work was first published in 1856 when Kerner was 70.

study of natural phenomena framed by a scientific method. Most importantly, Newton foregrounded the idea of nature as uniquely mysterious and governed by complex, divinely inspired forces that were discoverable through experimentation. Newton's own personal investigations often involved placing himself at physical risk, including introducing a large needle to his eyeball in an attempt to demonstrate the nature of colours. As Michael White has written: 'As a result, by nearly causing permanent blindness, he came close to destroying his scientific career almost before it had begun'.[2] Newton's entire investigative project centred upon an attempt to *physically* understand the component forces of the universe using all the data at hand. Despite the image of Newton as the rational herald of a new approach to exploring nature, his methodology and range of interests were varied, exotic and owed a great debt to the ancient and mysterious practice of alchemy.

It's perhaps unsurprising that others applied Newton's ideas in ways that were also highly imaginative. In 1766, 39 years after the death of Newton, Franz Anton Mesmer completed his doctoral dissertation. Its title was distinctly Newtonian: *De Planetarium Influxu in Corpus Humanum*—On the Influence of the Planets Upon the Human Body. Mesmer suggested that, if Newton's theories had indeed harmonized celestial and terrestrial mechanics, then the human body could not remain unaffected. He went on to describe 'tides' that might exist within the human subject that could be understood and manipulated. In 1768, Mesmer moved to Vienna and began experimenting with magnets. His social status rose considerably, and he began describing himself as a doctor. In 1774, Mesmer claimed to have produced a 'tidal effect' in a female patient called Francisca Osterlin. This was achieved—supposedly—by Osterlin swallowing a preparation that contained iron, with Mesmer attaching magnets to her body that he then moved. His theory was that the body possessed poles, as the earth does, and magnetic energy within the body could be manipulated to create a 'crisis', something like a seizure. The result of this would be improved health via a rebalancing of natural forces.

Despite patients reporting successful responses to treatment, Mesmer soon abandoned the use of physical magnets and focused instead on the development of a grander, pseudo-Newtonian theory. Magnetism, he suggested, existed in significant amounts in certain individuals and could be

[2] Ibid., page 61.

conducted and transmitted by them. One of the most effective conductors of this power was—of course—Franz Anton Mesmer himself. He theorized that a fine, invisible fluid surrounded all objects and explained all universal activity, from the planetary to the microscopic (a world presented in print in Robert Hooke's astonishing work of 1665, *Micrographia,* along with the author's beautiful illustrations).[3] Mesmer termed his newly discovered force 'Animal Magnetism'. Mesmer's reasoning upon this force was described by his biographer, Justinus Kerner:

> There must exist a power which permeates the universe, and binds together all the bodies upon earth, and it must be possible for man to bring this influence under his command.[4]

Scandal pursued Mesmer, leading to his relocating to Paris in 1778 where his patients consisted largely of the wealthy and influential. He developed an elaborate series of mechanisms for 'conducting' his power, including the 'baquet', a circular bath filled with water with iron rods emerging from it. Patients, joined together or holding hands, would supposedly amplify the magnetic force, passing it on to each other.[5] Some would report wild changes to consciousness and the production of trance states. Mesmer taught his theory and practices to students, creating a devoted following. Kerner described Mesmer's twenty-seven aphorisms in his biography *Franz Anton Mesmer: The Discoverer of Animal Magnetism.* His tone throughout the work is that of an awed disciple. The Aphorisms contain the following description of the invisible fluid's tidal power:

> IV. From this activity spring more or less general alternating operations which may be compared to ebb and flow.
> V. This ebb and flow are more or less general, more or less complex according to the nature of the origin which has called them forth.[6]

---

[3] The full title of Hooke's work is: *Micrographia or Some Physiological Descriptions of Minute Bodies Made by Magnifying Glasses. With Observations and Inquiries Thereupon.* It was published by the Royal Society.

[4] Ibid., page 34.

[5] Such activity, with individuals holding hands in a circle to supposedly conduct or share energy, created the basis for the 'séance', a term later used in spiritualist practice.

[6] Ibid., page 61.

Mesmer continually framed his 'discovery' in Newtonian terms, with references to 'mechanical laws', 'unity', 'vacuums' and 'influence'. He used the *Aphorisms*, along with the enthusiastic endorsements of well-to-do patients, in an attempt to gain medical recognition for his discoveries but without success. In 1784, a French royal commission was set up to establish the veracity of his claims. The panel of investigators included Benjamin Franklin, who was no stranger to the investigation of invisible forces in nature. It was concluded that Mesmer's treatments were scientifically valueless, and that any beneficial effect took place in the imagination of the patient. Yet Mesmer's practices continued to gain in popularity, despite the opinions of the establishment. Kerner's description of his hero gives a powerful sense of the esteem in which his devotees continued to hold him, and the power that Mesmer appeared to embody:

> Through this life, in the bosom of free nature, he appears even whilst still a child to have drawn towards himself a natural power unpossessed by the dwellers by the fire-side, a power which appears to delight to flow into those who maintain a many-sided intercourse and struggle with nature.[7]

To his followers, Mesmer had access to a unique sort of power, drawn not from religion, but from the natural world around him. Early in his career in Germany, Mesmer had debunked a series of exorcisms conducted by an Austrian priest. Mesmer concluded that the priest's power to heal came not from faith but from a surfeit of animal magnetism in his body. Mesmer's explanation was part of a progressive movement away from religious authority and towards a concept of accessible power embodied in nature. Despite his claim that such power was only *in abundance* in certain people, he believed that all humans possessed it to some degree. In this sense, Mesmerism was strangely democratic and reflective of broader political challenges to the dominant social and religious narratives of the late-eighteenth century. The foregrounding of ideas of natural force was derived in part from German 'Nature Philosophy' which sought to explain the world in terms of a unifying spirit or active power. This concept bound all things together in a way that was profoundly threatening to the doctrines of the Church since it located 'natural' force within the bodies of all living entities, eschewing centuries-old ideas of ordered and divinely sanctioned social structure.

---

[7] Ibid., page 32.

Mesmer's ideas also owed a debt to the works of the scientist-turned-mystic Emanuel Swedenborg. He was born Emanuel Swedberg in 1688 in Stockholm. His father, Jesper Swedberg, was a pastor who became Bishop of Skara. Emanuel was raised in an intellectually febrile atmosphere and learned Greek and Latin, adding Hebrew during his studies at Uppsala University. Eric Benzelius, his brother-in-law, was a librarian with an interest in philosophy who nurtured his scholarly interests. In 1710, Swedenborg travelled to London where he attended Newton's lectures at the Royal Society and met Edmond Halley and John Flamsteed. By his mid-twenties, he had designed prototypes for a flying machine and for a submarine, as well as publishing a volume of poetry. On his return to Sweden, he attempted to convince King Charles XII to fund the building of an observatory in the north of the country. The king refused but appointed him to the role of extraordinary assessor on the Swedish Board of Mines, recognizing Swedenborg's brilliance across a range of disciplines, including metallurgy. Despite his new duties, he still managed to edit a new scientific journal, *Daedalus Hyperboreus* (The Northern Daedalus). For this new periodical he produced an influential new work *On Tremulation*, in 1719. It's a theory of 'vital forces' that attempts to describe what we now call the nervous system, with vibrations passing from the brain and activating the rest of the body. To make his point, Swedenborg used an artistic analogy:

> the membranes, over which tremulation flows and which carry the motion into the cranium and over the whole osseous system, are of the same quality as a musical chord.[8]

The medium for tremulation—because all waves require one (this will get problematic later)—is 'a fluid...an animal spirit; this fluid has thus flowed out of the medulla through the nerves into the membranes'.[9] For Swedenborg, this fluid circulates around the body in a similar way to blood: 'so also does the nervous fluid possess its own glandules, ventricles and vessels whence it is distilled in the brain'.[10] It's a remarkable attempt to understand and map the nervous system. As we will also see in our

---

[8] Emanuel Swedenborg, trans. C TH Odhner, *On Tremulation* (Boston: MNCU, 1899) page 29.
[9] Ibid., page 31.
[10] Ibid.

exploration of the uses of electricity, this mysterious 'nerve power' was initially conceptualized as a kind of fluid. Swedenborg and Mesmer both viewed the nexus in the body to be a material system of refined liquid that might be manipulated and balanced to aid the health of the individual. Their thinking was undoubtedly influenced by the Hippocratic idea of 'humours' or liquids in the body that became unbalanced creating specific types of temperament in each individual. This was not entirely disbelieved until the mid-nineteenth century, yet the idea of a network that joined brain and body and described the carriage of impulses *within* the body was growing in plausibility in the decades before the French Revolution.

As the son of a prominent theologian, it's not surprising that Swedenborg also addressed metaphysical questions. In his work of 1734, the Newton-inspired *The Principia*, he attempted to address how the infinite might create the finite:

> in the producing cause there was something of a will that it should be produced, something of an active quality that produced it: and something of an intelligent nature determining that it should be produced in such a manner...in a word, something infinitely intelligent.[11]

The key for Swedenborg was the existence of patterns or vortices of force, nebulae that turned and generated matter yet always remained connected. It was an astonishing idea that, from a twenty-first-century perspective, appears to look both backwards and forwards. Back to the Renaissance and to neo-Platonic ideas of a sequence of circular worlds emanating from a 'divine mind' or 'nous', and forwards, albeit in a remote way, to unifying concepts in modern physics such as string and twistor theories. But where was God in this extraordinary picture? Was He within the universe, or outside it?

Both Swedenborg and Mesmer were troubled by ideas of the duality of mind and matter that had prevailed since the early seventeenth century via the philosophy of René Descartes. Wrestling with this problem, some argued, cost Swedenborg his sanity as he wrote increasingly wildly about mystical and interior experiences. As Wouter J. Hanegraff has stated:

---

[11] Emanuel Swedenborg, trans. J. Randel and E. Tansley, *The Principia or the First Principles of Natural Things* (London: The Swedenborg Society, 1912) page xxxvii.

Trained in the Cartesian philosophy of his day, with its strict separation between matter and spirit, he experienced a deep religious crisis in 1744: forced to admit to himself that his scientific explorations led him to the 'abyss' of pure materialism, he prayed to God and was granted a vision of Christ.[12]

Swedenborg experimented with a series of trance states and practised hypnagogia[13] as a means to converse with spirits and angels. Interestingly, in the Swedenborgian landscape of the afterlife, angels had once been human and continued to lead lives in the next world that we might regard as somewhat workaday. Indeed, the visionary himself continued to lead an active public life whilst keeping his 'dream diaries' and did not retreat into hermetic isolation. His influence on the ideas of the eighteenth and nineteenth centuries cannot be ignored, and we will encounter him again in our explorations of telepathy and telekinesis. His concept of tremulation was an insightful and rigorous attempt to explain the complexity of the nervous system and to unravel a mystery that might connect visible and invisible worlds. The philosopher Immanuel Kant brutally dismissed his later metaphysical speculation in his work of 1766, *Dreams of a Spirit Seer*:

(It is) eight volumes quarto full of nonsense. He puts them before the world as a new revelation under the title of '*Arcana Coelestia*'.[14]

Kant appears to view the older Swedenborg as a shadow of his former self, but, like Newton, Swedenborg's love of the scientific *and* the mysterious were part of the same thing: the search for an alchemical pattern that was working in both spirit and matter. As Gary Lachman has written of Swedenborg's theory of correspondences, the material world 'is on the lowest rung of a kind of ladder of worlds, with higher ones reaching above us into realms beyond the physical'.[15] The activities of 'infinite' or divine

---

[12] Wouter J. Hanegraaf, *Western Esotericism A Guide for the Perplexed* (London: Bloomsbury, 2013) page 37.
[13] The transitional state between waking and sleeping sometimes connected to 'lucid dreaming' and aspects of mystical experience.
[14] Immanuel Kant, trans. E F Goerwitz and Intro by F Sewall, *Dreams of a Spirit Seer Illustrated by Dreams of Metaphysics* (London: Sonnenschein and Co., 1900) page 101.
[15] Gary Lachman, *Introducing Swedenborg: Correspondences* (London: The Swedenborg Society, 2021) page 7.

agencies created, eventually, the world that we can perceive. Matter is in a constant state of transformation from one form to another.

It's easy to deride Swedenborg's vital force and Mesmer's theory of animal magnetism, yet they are both, in their own way (highly fanciful) descendants of Newton's gravity. It was reasonable to assume that, if celestial and terrestrial mechanics acted in unity, then human bodies must be affected in some way by such a colossal, universal force. The idea of a superfine fluid that binds all things and connects two worlds is also reminiscent of ancient ideas about the cosmos. The power of water was unleashed in new ways during the eighteenth century with the development of steam. Newcomen's Atmospheric Engine was invented in 1712—during Newton's lifetime—and condensed steam into a cylinder to produce mechanical work. This might be seen as a form of transmutation on an almost alchemical scale, with water becoming a vapour with sufficient power to move pistons and turn cogs. Previously, factories had made use of water via rivers, using wheels to power considerable mechanical tasks. But this new method controlled, intensified and internalized it. It was, therefore, hardly irrational to view water as possessing mysterious and powerful qualities. Steam replicated the power of a river in miniature, and, over the following decades, steam technology was applied to a range of processes. Like all technologies, steam-powered mechanisms became smaller, more efficient and easier to produce. By 1830, Manchester had hundreds of factories powered by steam technology, and some viewed this as a kind of continuous, secular miracle: just as water became steam, so the factories it powered turned cotton into clothing. Such processes, which quickly transformed all aspects of life, must have seemed like a frightening kind of alchemical force to those unaware of the scientific processes in play. The world was becoming a place of constant transmutation.

I have digressed here to show that it was not necessarily foolish to seek variations of revelatory power in the substance and structure of everyday phenomena. It was also part of the complex and shifting conceptual landscape of the eighteenth century. Mesmer's later methods were deeply dubious, and his propensity to share his healing with the wealthy exclusively does not make him particularly sympathetic to a modern reader. Yet the forces of nature were already being harnessed and conducted by new kinds of enterprising enthusiasts across western Europe. Transformation for profit was the aim, an intention certainly shared by the alchemists of the Early Modern period. Most significantly, hidden forces (real and imagined) were being discovered and manipulated by those beyond—and some way beyond—the social elites of Europe. These natural powers had uses that did not need the approval of monarchs or bishops, they could have immediate and direct applications in areas of wealth-production and well-being. This, as we shall see, was especially true in the discovery and development of electrical *current*. It might be invisible, but its effects were very real and potentially available to all. The new powers in play were powers for the people.

Mesmer's strange theory of a pervasive and magnetic fluid found an unusual ally in the next generation of European intellectuals in the form of Arthur Schopenhauer. Born in Danzig (now Gdansk) in 1788, Schopenhauer was from a family of wealthy merchants who supported the ideals of the French Revolution. Although trained to join the family business, Schopenhauer was deeply drawn to philosophy. He derided the then-fashionable ideas of George Hegel and became one of the first western scholars of Indian literature. He was particularly drawn to the Vedas and the ancient preoccupation with complex concepts of illusion and underlying reality. This became a propulsive element in his thinking.

It's important to understand that Schopenhauer's background was extremely practical. His family profited from developments in technology and he travelled extensively—including to Britain—to learn about the growing networks of production and distribution. His interest in the work of Kant began when he read the *Critique of Pure Reason* and believed that the philosopher had stumbled upon an astonishing but frustrating truth. In his attempts to reconcile the philosophical schools of Rationalism and Empiricism, Kant produced a grand theory: Transcendental Idealism. This was Newtonian in its scope and ambition, a philosophical theory of everything for its time. Kant believed that the world appears to us as a series of phenomena framed by causality, space and time. Our senses receive the world as experience, and our brains decode, structure and make sense of it. The *true* nature of it—the *noumenon or 'thing in itself'* as he termed it—is invisible and inaccessible to us since it cannot be experienced directly. And yet it was responsible for *everything*, underpinning and maintaining all things. The Newtonian influence is quite clear. Yet the grand project that Kant hoped would reveal the very building blocks of existence resulted in a dead end, and a mystery he simply could not resolve. As he wrote somewhat mournfully in the second edition of the *Critique of Pure Reason*, published in 1787:

> Even if we could bring this intuition of ours to the highest degree of distinctness we would not thereby come any closer to the constitution of objects in themselves. For in any case we would still completely cognize only our own way of intuiting, i.e., our sensibility, and this always only under the conditions originally depending on the subject, space and time; what the objects may be in themselves would still never be known through the most enlightened cognition of their appearance, which alone is given to us.[16]

Kant's influence has continued across the centuries in many forms: the Logical Positivism of Wittgenstein and the General Theory of Relativity of Einstein. Yet his ambition was, like Newton's, strangely alchemical. It was an attempt to understand the connection between our perceptual powers and the world itself, and the near-magical act of transmutation that turns raw data into consciousness. Schopenhauer could not accept his hero's failure in his great project. In short, human knowledge of the noumenon could not be permitted to remain a mystery. Schopenhauer solved the problem via what he called 'the secret door'. His major work *The World as Will and Representation*, first published in 1818, suggested there was a solution to Kant's mystery, a way to experience the noumenon—through our awareness of our own bodies. In brief, we know the world—including ourselves—through sense data that we understand as phenomena—and yet we are also aware of an inner existence, a force we cannot control that is sustaining us via our breathing, heartbeat and blood flow, and the unconscious activity of the brain. By inner meditation, we apprehend—to some degree—this mysterious force. Schopenhauer, somewhat problematically, termed it 'will'. He identified this will as the source of materiality, existence and consciousness. It underpins all things in the universe—sentient and non-sentient. As he wrote in *The World as Will*: 'Therefore, the hungry wolf buries its teeth in the flesh of the deer with the same necessity with which the stone falls to the ground'.[17] The will cannot be judged—it simply *is*, 'the Will to Be'—a term Friedrich Nietzsche would later elaborate upon and eventually transform into something else entirely. To Schopenhauer, the will is our world. When we encounter this bleak iteration of a transcendent force, our only escape from it—according to the philosopher—is in meditation and the contemplation of art. This has made

---

[16] Immanuel Kant, trans and ed P Guyer and A W Wood, *Critique of Pure Reason* (Cambridge: CUP, 1998) page 185.

[17] Arthur Schopenhauer, trans. E F J Payne, *The World as Will and Representation*, Volume One (New York: Dover, 1969) page 404.

Schopenhauer's writing hugely influential upon artists and writers, including Richard Wagner, Thomas Hardy and Samuel Beckett. There is little that can be done to avoid the universal and remorseless will, as Schopenhauer wrote:

> No attained object of willing can give a satisfaction that lasts and no longer declines; but it is always like the alms thrown to a beggar, which reprieves him today so that his misery may be prolonged till tomorrow. Therefore, so long as our consciousness is filled by our will, so long as we are given up to the throng of desires with its constant hopes and fears, so long as we are the subjects of willing, we never obtain lasting happiness or peace.[18]

We might be forgiven for concluding that Schopenhauer's dour interpretation of the will in nature owed much to his own cultural background—a rather chilly variation of Protestantism suffused with an aggressive work ethic. His family history contained a concerning number of suicides. In some respects, Schopenhauer's concept of a disinterested will is the negative version of both late-Swedenborgian concepts of a transcendent, divine force and Nature Philosophy's vision of a unified living world. But his conclusions also show the powerful influence of Hindu and Buddhist thought on his world view.

Despite directly addressing the problem posed by Kant's *Critique*, Schopenhauer's work was largely ignored in his own time. This can be at least partially attributed to his difficult personality, misanthropy and stubborn refusal to stop scheduling his lectures in Berlin at precisely the same time as the far-more-popular Hegel's. According to David E. Cartwright:

> Almost ten years to the date from the appearance of The World as Will and Representation, Schopenhauer contacted his publisher discreetly as to whether a second edition was needed…despite the fact that he had been assiduously developing material to clarify, augment, and extend his ideas into new areas of enquiry…Of the original run of 750 books, 150 remained on the shelves.[19]

Schopenhauer's response to the neglect of his work was to gather together the notes from several of his unheard lectures and publish them as a new work entitled *On the Will in Nature*. The first edition appeared in

---

[18] Ibid., page 196.
[19] David E. Cartwright, *Schopenhauer: A Biography* (Cambridge: CUP, 2013) page 434.

1854, with a second being presented posthumously in 1867. In the latter version, he addressed the work of Mesmer. This was surprising, since he had little time for Mesmer's biographer and disciple, the aforementioned Justinus Kerner. In 1855, Kerner had published a work about a 'clairvoyant somnambulist' called Friederike Hauffe, known as 'The Seeress of Prevost'. Schopenhauer read the book and, despite admiring Kerner's detailed descriptions of the phenomena supposedly generated by Hauffe, dismissed the author's insistence on an objective, 'spirit reality' that existed independent of the will. As Schopenhauer wrote with characteristic intellectual contempt:

> In such a stupid world-order I would certainly not like to become so blessed, but would sooner saunter around in order to be able to sneer at it undisturbed.[20]

Yet Schopenhauer wasn't entirely dismissive of mysterious phenomena. Indeed, he was convinced that animal magnetism was a force—like Newtonian gravity—that attracts all beings in the universe as a manifestation of the will. As he wrote:

> Further, because the will manifests itself in Animal Magnetism downright as the thing-in-itself, we see the principium individuationis [principle of individuation] (space and time), which belongs to mere phenomena, at once annulled: its limits which separate individuals from one another, are destroyed.[21]

Schopenhauer was therefore able to dismiss concepts of spiritual and miraculous revelation, replacing them in his metaphysical system with a new explanation: the sudden, transcendent apprehension of the will by humans in exceptional circumstances. At such points of crisis, he believed, the frames of space, time and causality disappear. The appearance—or 'representation'—of the gravity-like will that holds all things in a perceived order—is momentarily cancelled. Poltergeists, clairvoyant experiences and spiritual trances were unfiltered experiential manifestations of the mysterious force.

---

[20] Ibid., qtd in Cartwright page 439, trans E F J Payne.
[21] Arthur Schopenhauer, trans. J. Frauenstadt, *Two Essays by Arthur Schopenhauer* (London: George Bell and Sons) page 332.

It's fanciful stuff and, like most of Schopenhauer's thought of the time, it went largely unnoticed. And yet it was remarkable in its attempt to construct a naturalistic explanation for supposedly 'supernatural' experiences and drew the western philosophical tradition somewhat closer to far older eastern concepts regarding unexplained incursions into the quotidian. David E. Cartwright writes of Schopenhauer's selective credulity:

> He tended to accept magnetism and other paranormal phenomena at face value and with the same nonskeptical attitude he expressed in his philosophy. Although he was generally suspicious about the motivations of people he encountered in his everyday life…his attitude was much different towards anything that he thought could help corroborate his philosophy.[22]

For many philosophers, Schopenhauer's work was an elaborate and beautifully constructed misstep: the point of the Kantian noumenon is that it is inherently mysterious and *cannot* be directly experienced. The door to total understanding is closed. But it's possible to view the imaginative works of Mesmer, Swedenborg and Schopenhauer as attempts to solve a mystery that had begun with Newton: what substance or force constitutes and shapes the phenomena of the universe and to what extent is that substance or force present in us? In the next chapter, I'll consider the influence of Mesmer's ideas, as his followers focused increasingly upon sharing his techniques beyond the social elites of the nineteenth century, whilst also developing Mesmerism as a force that worked, predominantly, upon the mind. Despite some astonishing success stories, Mesmer's strange blend of magic and science failed to throw off its association with showmanship and fraud until it was reformulated under a new name decades later: hypnotism.

---

[22] Cartwright, page 443.

CHAPTER 4

Forces in the Mind

Abstract In this chapter, I examine the fate of Mesmer's ideas with a particular focus on the Victorian physician John Elliotson. I consider Elliotson's refinement of Mesmer's approach and the consequent arguments about the nature and legitimacy of Mesmerism. I then go on to look at the interpretation of Elliotson's research by James Braid, who concluded mesmerism was, largely, a mental process. I proceed to consider Victorian anxieties about the manipulation of mental forces, as evidenced in the stage play *The Bells*.

Keywords Mesmerism • Hypnotism • Trance • John Elliotson • James Braid • Henry Irving • Bram Stoker • Elizabeth Okey

> It is the character of a humane man to be anxious that all promises of benefit to his fellow creatures may be fulfilled; that every alleged means of curing disease may turn out, not a fallacy, but a reality. (John Elliotson, *Numerous Cases of Surgical Operations Without Pain in the Mesmeric State*)[1]

Franz Anton Mesmer's career had begun with a sincere and admirable attempt to reconcile the effects of the vast, cosmic forces described by Newton with the activity of the human body. Mesmer's initial

---

[1] Ibid., page 55.

investigations had been revealing and accounts of his practice detailed some notable successes, as well as a number of failures and scandals. His enjoyment of the theatrical aspects of his séances and propensity to work with wealthy clients led to his being discredited, yet his practice created a number of committed followers who would, over time, modify and develop the elements of mesmerism. In London, mesmerism was largely viewed as a fringe activity promoted by the French. It was certainly looked upon with disdain by many in the British medical establishment. In 1842, three years after successfully removing mesmerism and its principal British advocate from University College Hospital (UCH), London, the founder of *The Lancet* Thomas Wakley wrote:

> Mesmerism is too gross a humbug to admit of any further serious notices. We regard its abettors as quacks and impostors. They ought to be hooted out of professional society. Any practitioner who sends a patient afflicted with any disease to consult a mesmeric quack ought to be without patients for the rest of his days.[2]

Wakley had founded the journal in 1823 as part of a movement to encourage reputable accounts of medical processes, to expose malpractice and to encourage debate within the field. He became MP for Finsbury in 1835 and pursued a progressive agenda with a focus on political reform. He was known as a champion of the London poor and held his parliamentary seat until 1852. His dislike for mesmerism should be viewed as part of a large and complex debate about the reform of nineteenth-century medical practice. The struggle over the legitimacy of scientific processes, and the arguments for what constituted reputable practice, would continue beyond the mid-nineteenth century. For Wakley, neutralizing the threat of mesmerism—in his opinion a pseudo-science conducted with suspicious levels of showmanship—was an imperative and part of a broader struggle that involved the control and direction of University College Hospital, London. UCH's most prominent teacher was John Elliotson.

In 1834, Elliotson was appointed as senior physician at University College Hospital. He was 43 years old, and his career—like Wakley's—had been inspired by reformist ideas. Both men cared deeply for the health of the poor and were committed to improving the conditions in which medical practice was undertaken. Elliotson was open to experimentation and

---

[2] *The Lancet*, 29th October 1842.

happy to test treatments based on accounts of their success, even if their source might be considered pseudo-scientific. In 1843, he defended his seven-year campaign in support of mesmerism in the following way:

> I have never speculated but have always devoted myself to the observation of facts so that, whatever I have advanced, I have seen ultimately established. I appeal to my writings on quinine, hydrocyanic acid, iron…and the use of the ear in ascertaining the state of the heart.[3]

The last statement refers to Elliotson's championing of the stethoscope—invented in France by René Laennec in 1816—but rarely used in English medical practice. In 1829, Elliotson had been impressed by a demonstration of animal magnetism undertaken by the Irish physician Richard Chenevix. In his work of 1835, *Human Physiology*,[4] he dedicated a lengthy section to the discussion of mesmeric practice. Elliotson had witnessed a range of phenomena that he considered genuine but could not entirely explain. As he wrote of the experience:

> The patient becomes insensible to all around, but may have the inward senses augmented as in common ecstasis – may well sing for the first time in his life, and talk so unguardedly as to disclose secrets.[5]

Elliotson also described some of the more extraordinary feats reported by those 'magnetized' during mesmeric practice, phenomena we might now describe as ESP: 'to discover a person in the next room though a wall intervene…as well as learn the thoughts of persons present'.[6] Although never witnessing anything he considered 'supernatural' in person, Elliotson was open to the possibilities of mesmerism. At the same time, he was deeply critical of Franz Anton Mesmer, writing of him:

> He travelled and performed many great cures, and often failed: was praised and deservedly abused, for he adopted the course of all quacks, whether

---

[3] John Elliotson, *Numerous Cases of Surgical Operations Without Pain in The Mesmeric State* (Philadelphia: Lea and Blanchard, 1843) page 51.

[4] The book began life as the translation of a work by physiologist Johann Friedrich Blumenbach but Elliotson's significant contributions over time led to it being presented as a self-authored work.

[5] John Elliotson, *Human Physiology* (London: Longman, Rees, Orme, Brown, Green and Longman, 1835) page 660.

[6] Ibid., page 661.

regular or irregular practitioners. He depreciated others, affected mystery and extolled himself. He insisted there was but one health, one disease and one remedy, which remedy of course, he had discovered.[7]

But Elliotson believed that Mesmer had discovered *something* and supported the idea of 'the existence of a fluid pervading all living and inanimate matter',[8] a rather Schopenhauerian idea that permitted the human subject, and the medical practitioner, access to the life-sustaining processes within the body. For Elliotson it might be possible for forces within the body to be manipulated in some way by external touch and manipulation. As he wrote in *Human Physiology*: 'Brahmins and Parsis and the Jesuit missionaries inform us that the practice of curing disease by the imposition of hands has pervaded in China for many years'.[9]

To demonstrate his lack of credulity, Elliotson describes a magic trick he'd seen practised in London, during which a supposedly clairvoyant boy was able to predict objects held up out of sight by decoding verbal signals given by his father. In the demonstrations of mesmerism presented by Chevenix, Elliotson insisted in speaking French to ensure their subject—a young Irish woman who had lost the use of an arm—was unable to obtain any information from their discussion. The woman was placed 'deep in a trance' and Chevenix pointed towards her arm. 'He told her to raise it: She could scarcely move it'. After a few 'transverse movements' by Chevenix, the woman 'declared the stiffness and uneasiness were gone and she moved it as well as the other'.[10] Elliotson observed that—although Chevinix's results were far from consistent—they had a value and proved 'such a power of mesmerism exists'.[11] A French commission of 1826 had come to rather different conclusions to that set up before the French Revolution, and concluded that there were 'sometimes decided effects'.[12] In 1837, Elliotson got the opportunity to watch a mesmerist up close. His name was Jules Denis Dupotet, who styled himself as 'Baron' Dupotet. He was one of a number of disciples of Mesmer who continued to refine their teacher's techniques after his death. But Mesmer's questionable reputation had made it difficult for him to obtain 'respectable' opportunities

[7] Ibid., page 663.
[8] Ibid., page 662.
[9] Ibid., page 667.
[10] Ibid., page 681.
[11] Ibid.
[12] Ibid., page 688.

to practice in London. Elliotson, however, gave him such an opportunity, inviting him to demonstrate his techniques on a number of patients at UCH in the summer of 1837. Although clearly convinced by some aspects of mesmerism, Elliotson was still rather sceptical. However, Dupotet's work with supposedly 'incurable' patients soon convinced him. Thomas Orton, a stable hand who had been diagnosed as an 'epileptic'—a term that covered a range of symptoms at the time—had been treated with the conventional remedies of leeches and enemas with no discernible improvement in his condition. Dupotet succeeded in putting him into a trance state via a series of gestures, 'passes' and by staring into his eyes. Every day for three months, Dupotet repeated the ritual to the point where Orton's seizures ceased completely. He was discharged from UCH in 1838. Dupotet enjoyed considerable success in the treatment of patients with epileptic symptoms, yet the theatricality of his presentation soon grew irksome to a number of the senior faculty at UCH—including Robert Liston—the chief surgeon. During Elliotson's absence overseas in the autumn of 1837, Dupotet's patients were taken from him, and he was asked to leave the hospital. Liston's friendship with Thomas Wakley was influential in the marginalization of mesmeric practice. Yet, as Elliotson pointed out, he had always operated within the accepted rules of scientific method. In *The Zoist*—the journal Elliotson subsequently founded—he documented his failures as well as his successes, and readily admitted that the techniques did not generate regular, predictable results or enjoy universal success. He remained true to his observations of 1835 (and the French commission of 1826) regarding the consistency of results:

> Whatever the phenomena developed in the mesmeric sleep-waking (trance) I have observed no rule for either the period or order of their occurrence, or for their duration, nor any great relation among them.[13]

Despite the opposition of many of his colleagues, Elliotson continued to develop the techniques he had seen Dupotet employ at UCH. He focused primarily on one subject—a young woman called Elizabeth Okey.

Elizabeth Okey had been admitted to UCH in the spring of 1836 at the age of 16. During a fall she had suffered a head injury and endured regular seizures. Her younger sister Jane had already been admitted to

---

[13] *The Zoist: A Journal of Cerebral Physiology and Mesmerism, And Their Applications To Human Welfare*, Vol. 2 page 201.

UCH on account of epilepsy, and Elizabeth's symptoms mirrored those of her sister in some respects. Elizabeth was one of the first patients to be treated by Dupotet. As Wendy Moore has written:

> Elliotson exposed her (Elizabeth) to his usual battery of remedies. She was bled repeatedly by the lancet and with leeches, cupped, blistered and administered a large daily dose of silver nitrate…She often had as much as ten ounces of blood (roughly half a pint or 284 ml) removed at a time and this was repeated every two or three days.[14]

The positive effects of these then-conventional treatments were negligible, but Elizabeth proved highly receptive to mesmerism, falling into trance states quickly and demonstrating a range of phenomena that were strangely performative. Despite appearing shy and quiet in her waking hours, in trance Elizabeth became animated and vocal, speaking in a range of different accents and entertaining the male audience with anecdotes. Most remarkably, Elizabeth appeared impervious to pain. As Elliotson later wrote of the experiments performed during the trance state:

> I had myself witnessed five years previously the introduction of a seton (needle) into the back of the neck of Elizabeth Okey without sensation…For a length of time she had perfect loss of the sense of touch – anaesthesia – in her ecstatic delirium. She could do nothing unless she saw it nor, 'till she acquired the habit, could she walk without looking at her feet. She used to take red-hot coals out of the fire and wonder, as she held them, why other people cried out and desired her to put them down, and why her hands became blistered.[15]

A similarly dangerous range of experiments were attempted on Elizabeth's sister Jane with equally shocking results: 'there was always insensibility of touch and cupping and the severest blistering were perfectly un-noticed'.[16]

Throughout this period of experimentation on the Okey sisters, Elliotson continued to believe in the presence of a vital fluid within the

---

[14] Wendy Moore, *The Mesmerist: The Society Doctor Who Held Victorian London Spellbound* (London: Weidenfeld and Nicolson, 2018) page 83. The author gives a detailed and highly readable account of Elliotson's work and his rivalries.

[15] John Elliotson, *Numerous Cases*, page 40.

[16] Ibid.

body that could be manipulated by the use of magnetized objects. This approach was similar to Mesmer's early experiments, and it was not viewed positively by Liston or Wakley. In August 1838, Wakley invited Elliotson and the Okey sisters to his home at 35 Bedford Square in Bloomsbury. A series of tests were undertaken, with a piece of magnetized nickel applied to Elizabeth's face. A seizure followed, which Elliotson regarded as a clear validation of his theories. However, further experiments led to a variety of different responses from the sisters, and Wakley concluded that the effects of mesmerism were not detectable in the water or upon any of the metal objects that had been 'magnetized'. Wakley undertook his own experiments, published the results and used them to muster opposition within UCH to Elliotson's work. At the same time, a number of hostile articles in *The Lancet* denounced Elliotson as the victim of a sustained fraud by the Okeys. On 27 December 1838, the Council of University College banned the practice of mesmerism within the hospital. Elliotson resigned shortly after.

Far from being cowed by his ejection from UCH, Elliotson rallied and set up a practice at his home in Conduit Street, London. With the Okey sisters asked to leave the hospital, he offered them a room in the servants' quarters of his house. Although he initially suffered a considerable decline in income from the loss of his private clients, interest in Elliotson's cases continued to grow and attendance at the improvised clinic reflected this fascination. Despite scandalous accounts of the nature of the investigations that did the rounds in the popular press, Elliotson's social status recovered in the years following his expulsion by Wakley. In 1843, he offered a robust assertion of his medical reputation and defended the Okeys against all accusations of fraud, turning on Wakley in a stunning ad hominem attack:

> Mr Wakley… is a father, and he should have some feeling for innocent young females who, though in the humbler walks of life, are not his inferiors in respectability…His day of triumph has passed, and his chief business now must be to extricate himself from the sad position into which he has fallen from having so overcunningly, hastily, and violently committed himself.[17]

The conclusion of the work is a passionate rebuke against the dominant medical institutions and a plea for open-mindedness. Elliotson reminds his

[17] Ibid., page 52.

peers of the initial scepticism regarding Harvey's theory of the circulation of the blood and warns against scientific complacency and arrogance:

> just as the course of the earth taught by Pythagoras had to be taught afresh by Copernicus, at the end of two thousand years, after being reviled and then again required all the powers of Newton for its demonstration, so the truth of nerves and sense and motion being distinct fell into contempt in the last century.[18]

Elliotson's capacity to practice may have been partially curtailed by Wakley, but he had trained enough students in his methods to ensure that mesmeric experimentation continued to grow at an astonishing rate across Britain and the empire. His methods for placing patients in 'the healing sleep' were passed on and cases of mesmerism's success were regularly documented. In 1842, a barrister called William Topham, who had researched Elliotson's methods, put them into practice on a farm worker from Nottingham called James Wombell. Wombell had injured a knee so severely that his leg required amputation. The chief surgeon of the district hospital contacted Topham who duly attended and placed Wombell in a trance. Before a stunned crowd, the surgeon removed the leg at the thigh while Wombell remained deep in a trance state. He showed no sign of discomfort. Despite the growing demand for mesmerism from patients, Wakley continued to attack the practice via *The Lancet* at every available opportunity. For him, and for many of his peers, mesmerism was a fraudulent, exotic and continental practice that could not be demonstrated in controlled conditions over time. But Elliotson had always conceded that the techniques did not work on every patient. To his mind, the differing levels of success in mesmerized subjects might be explained by the force of animal magnetism itself: it simply existed in greater abundance in certain individuals or was perhaps boosted by some sympathetic connection between the mesmerist and their subject. Elliotson's practice is noteworthy for its holistic approach and capacity to understand treatment and recovery as an ongoing process that differed with each patient. Just as Wakley found his approach questionable, so Elliotson could claim that the supposed 'tried and tested' methods of bloodletting and leeches were similarly dubious, with little compelling evidence of success to support their continued use. What mattered to Elliotson was the effectiveness of the

---

[18] Ibid., page 56.

treatment for the *individual*, not its repetition over a host of other patients with different symptoms and needs.

Across the British Empire in India, surgeon James Esdaile began experimenting with Elliotson's techniques. He placed his patients in a trance and was able to conduct a series of previously agonizing operations—mostly the removal of tumours—without causing the subjects any apparent pain. In 1845, at the request of the Deputy Governor of Bengal, Esdaile became chief surgeon at a hospital in Calcutta where he trained his fellow practitioners in mesmeric trance and treated large numbers of patients. Esdaile appears to have blended mesmeric practice with a far older Indian tradition known as 'Jar Phoonk' that placed the healer and the patient in continuous contact via touch and breathing. The emphasis on time spent with the patient and in creating a shared experience between healer and subject was distinctly different from the battery of invasive procedures familiar to the sick in Britain. Using the blended technique, Esdaile was able to prepare the patient for procedures that would previously have generated unbearable levels of pain. Far from challenging the variations of his practice that his followers developed, Elliotson approved of them, once again demonstrating his commitment to the success of the treatment over its provenance.

From 1843 to 1856, Elliotson edited *The Zoist: A Journal of Cerebral Physiology and Mesmerism and Their Applications to Human Welfare*. Elliotson's journal was a place to defend mesmerism and to document and discuss recent cases. In the editorial in the edition of January 1845, a fierce defence of 'Intellectual Freedom' is mounted:

> Persecution is the offspring of ignorance and superstition. Liberality of sentiment and tolerance are the companions of knowledge and freedom. A persecutor is a man who has never inquired into the foundations of his own opinions; he invades the domain of thought with a club, and bids his neighbour conform to his views and embrace his doctrines, or dread the result.[19]

Elliotson clearly felt the pain of his expulsion from UCH and the medical establishment deeply. Yet his practices continued to flourish in the ensuing decades and his supporters included Charles Dickens and Harriet Martineau. He proved that an individual could defy the establishment and survive as a practitioner. He also succeeded in sharing his practice with the

---

[19] *The Zoist*, page 442.

lay community on an unprecedented scale. Ultimately, mesmerism was superseded by a rival theory that owed much to the work of John Elliotson: hypnosis.

The Scottish surgeon James Braid had originally dismissed mesmerism on account of the supposed fraud detected by Wakley and documented in *The Lancet*, referring to it as 'a system of delusion or collusion'[20] but in 1841 he attended a demonstration by a Swiss mesmerist named Charles Lafontaine. In his work of 1843, *Neurypnology*, Braid recalled the moment, and his stunned inability to explain what he'd experienced:

> I saw one fact, the inability of a patient to open his eyelids, which arrested my attention...and I therefore instituted experiments to determine the question; and exhibited the results to the public in a few days after.[21]

Braid was convinced that mesmerism created a change of state in the subject, but opposed the core idea of Franz Anton Mesmer by denying that any kind of magnetic or other invisible force was being stimulated. He argued that what was at work was a psychological process in which the mind alone is acted upon and moved into a different state of consciousness. He succeeded in achieving many of the outcomes attributed to mesmerism without physical contact, but by getting the subject to concentrate on a specific object whilst addressing them in a certain way. It's Braid's version of mesmerism—hypnotism—which has survived as a therapeutic tool via the work of Freud and others. By divesting mesmerism of its performative elements and physical contact, along with the refutation of the idea of an occult force or fluid, Braid made the topic more acceptable to Victorian sensibilities:

> It will be observed, for reasons adduced, I have now entirely separated Hypnotism from Animal Magnetism. I consider it to be merely a simple, speedy, and certain mode of throwing the nervous system into a new condition, which may be rendered eminently available in the cure of certain disorders.[22]

Perhaps most importantly, Braid began the process of making hypnotism respectable and explicable to the world beyond the scientific elites. What

---

[20] James Braid, *Neurypnology* (London: George Redway, 1899) page 98.
[21] Ibid., page 116.
[22] Ibid., page 86.

troubled the moral sensibilities of concerned Victorians most was the idea of losing control whilst under hypnotic influence, and possibly undertaking behaviour that was not morally correct. It's interesting that, in the years following the popularization of hypnotism, there were a range of sinister portrayals of 'trance', 'somnambulism' and 'magnetic sleep' in literature and drama, often involving characters with mysterious and hidden 'other lives' such as Dr. Jekyll and Dorian Gray. Braid sought to address the concerns of an audience beyond that of Wakley's journal *The Lancet*:

> I am aware great prejudice has been raised against mesmerism, from the idea that it might be turned to immoral purposes. In respect to the Neuro-Hypnotic state, induced by the method explained in this treatise, I am quite certain that it deserves no such censure. I have proved by experiments, both in public and in private, that during the state of excitement, the judgment is sufficiently active to make the patients, if possible, even more fastidious as regards propriety of conduct, than in the waking condition; and from the state of rigidity and insensibility, they can be roused to a state of mobility, and exalted sensibility.[23]

However, hypnotism was never *entirely* reputable in Britain due to its earlier continental connections and ongoing association with magical performance. Hypnotism—an undeniably powerful psychological tool—remains mysterious in many ways since its processes and the nature of its effects upon the mind are still contested. It's depiction in the media remains controversial, with 'celebrity' hypnotists offering cures for a range of phobias and bad habits, and stage magicians entrancing audience members to indulge in a variety of bizarre activities, from acting like a chicken to taking part in a real-life version of a video game.[24] There is still an unsettling lack of clarity regarding hypnosis, and the same troubling overlap between therapeutic techniques and performative tropes experienced in the nineteenth century remains in place today.

The gestural 'passes' of the mesmerist and echoes of mysterious, near-priestly power sat ill with a scientific establishment that was, perhaps, too interested in demonstrating regularity of results and respectability of practice. Yet there's so much to admire about the pioneering work of John Elliotson. His belief in mesmerism was based on his experience of its efficacy in the alleviation of pain. It gave comfort to those who had been

[23] Ibid., page 91.
[24] Ibid., page 91

failed by orthodox medicine. From our viewpoint in the twenty first century, it's bloodletting and surgery without anaesthesia that seem barbaric. Elliotson treated patients as individuals and spent time with them. Their rehabilitation was an ongoing process. Undeniably, Elliotson's weakness was in his attachment to a small group of patients at UCH, and his desire to display the treatments publicly led to accusations that it was as much theatre as medicine. Braid believed that the effects of mesmeric treatment faded over time as the patient got used to it. If so, then Elliotson's work with the Okeys may have involved a degree of conscious or unconscious fraud. The lives of the sisters were probably better at UCH and in Conduit Street than they were at the family home. They had a vested interest in continuing to appear troubled and, at some point, Elliotson appears to have lost sight of the purpose of treatment: recovery and a return to normal life. It's possible that Elliotson's love of the mystery of mesmerism, and his desire to seek out ever-more exotic phenomena via its application, came to dominate and overwrite his scholarly and medical interest in the topic.

In 1846, Elliotson was chosen to deliver the Harveian Oration at the Royal College of Physicians, an address named after one of his heroes, William Harvey. Predictably, the decision was bitterly opposed by Wakley at *The Lancet*. Elliotson chose to talk about mesmerism. In the speech, he quoted Harvey:

> True philosophers, compelled by the love of truth and wisdom, never fancy themselves so wise and full of sense as not to yield to truth from any source and at all times: nor are they so narrow-minded as to believe any art or science has been handed down in such a state of perfection to us by our predecessors that nothing remains for future industry.[25]

Elliotson believed that mesmerism stimulated a powerful force that could then be manipulated to improve the health of the patient. This has, of course, proved to be incorrect, but there was something valuable at the heart of the strange practice that developed around mesmerism after Mesmer: it revealed the profound interconnectedness between mental and physical health and challenged the duality of mind and body that had existed in western philosophy for over two centuries. It also emphasized the uniqueness of the patient and their relationship to illness. Elliotson's

---

[25] John Elliotson, *The Harveian Oration* (London: H Bailliere, 1846) page 66.

animal magnetism was truly holistic. Perhaps it might be best to view it as a kind of metaphor for the bond between healer and patient. Just as Wakley and the Okey sisters might be equally respectable, so animal magnetism bound people together as equals in their capacity to access its power. It was an agent of *attraction* that could manifest itself in any person irrespective of rank or social caste. Not only that, but the techniques for manipulating it could be taught, learned and passed on to others. It offered a community of healers and a capacity to relieve pain that was no longer controlled by medical elites. Of course, there were dangers in that—the possibility of fraud and unregulated practice—but there was also something overwhelmingly progressive: a heightened interest in the complexity of well-being that has influenced and benefited the development of modern medicine.

Twenty-five years after Elliotson delivered the Harveian Oration, a little-known actor took to the stage in a new play at the half-empty Lyceum theatre in London. His name was Henry Irving, and, although he was ambitious to perform regularly in Shakespearean tragedy, he'd largely worked as a comedian in regional theatres. But he was cultivating a new sort of persona for performance: the sinister, conflicted gentleman with something to hide. The play was titled *The Bells*, and it had been adapted by a lawyer called Leopold Lewis from a French melodrama. Irving played the role of Mathias, a respectable burgomaster who is about to marry his daughter to a local gendarme. However, as the wedding draws near, Mathias is increasingly troubled by the sound of sleigh bells, and we learn that, to save the family home years before, he had murdered a merchant and stolen his money. In the play's startling third act, which became infamous for the levels of tension it developed, Mathias appears to pass into a nightmarish dream world where he's confronted by the officials of a spectral court. The robed President of the court questions Mathias about his past, but the burgomaster objects. The President summons a 'Mesmerist', but Mathias resists, to which the President replies: 'If you are innocent, why should you fear the Mesmerist? Because he can read the inmost secrets of your heart!'.[26] Mathias struggles, replying 'they only deceive people for the purpose of gaining money – they merely perform the tricks of conjurors'.[27] The Mesmerist undertakes no gestural passes at first, he merely

---

[26] Lewis, L., *Henry Irving and The Bells: Irving's Personal Script of the Play*, ed. D. Mayer (Manchester: MUP 1980), page 70.

[27] Ibid., page 71.

looks at him 'steadfastly' and Mathias duly falls into a deep sleep. It's at this point that the passes are performed 'behind him'—like a puppeteer perhaps—and Mathias 'succumbs under the influence'.[28] The next few minutes of stage time caused a sensation and ensured that Irving would perform this role around 800 times over the next 34 years. Under the Mesmerist's direction, Mathias doesn't just recount the murder, he acts it out, murdering the merchant again in mime, lifting the body and throwing it into a lime kiln to burn. Sentenced to hang by the President, Mathias shifts from the dream world back to the real one. His family break into his room on the morning of the wedding and find him clutching at his throat. He dies saying the line: 'Take the rope from my neck – take – the – rope – neck'.[29] It was Irving's capacity to intensely engage his audience that attracted the attention of his eventual manager, Bram Stoker, who relocated his life from Dublin to London to support the actor when he took on the lease of the Lyceum. Stoker remained Irving's confidante, manager and devoted fan until the actor's death in 1905. It's during this period that we first see the term 'to mesmerize' employed by theatre critics to describe outstanding performers.[30]

As we shall see in the next chapter, Mary Shelley's novel *Frankenstein* became a focus for fears about the nature of electricity. In a similar way, Henry Irving's performance in *The Bells* brilliantly distilled anxieties about mesmerism and hypnotism. It was over two decades since Braid's respectable defence of his hypnotic techniques, but the dominant view was that, in the best-case scenario, hypnotism was both deeply questionable and unacceptably continental. The good Victorian patriarch believed that he was fully in control of his body at all times, even if he could have little say in the matter of blood flow, respiration or the operations of his iris. Braid conceded, just as Elliotson had, that the techniques work better with some people than others, and there appeared to be no class or social categorization for those who were affected, or for those who could learn and practice the associated healing techniques. As Braid later wrote in *Neurypnology*, there was an element of mystery to the entire process and he harkened back to Newton's theory of gravity:

[28] Ibid.
[29] Ibid., page 76.
[30] This topic is explored more fully in W.D. King's *Henry Irving's Waterloo: Theatrical Engagements with Arthur Conan Doyle, George Bernard Shaw, Ellen Terry, Ellen Gordon Craig, Late-Victorian Culture, Assorted Ghosts, Old Men, War, and History* (Berkeley: UCP, 1993) pages 194–200.

As to the modus operandi we may never be able to account for that in a manner so as to satisfy all objections; but neither can we tell why the law of gravitation should act as experience has taught us it does act. Still, as our ignorance of the cause of gravitation acting as it is known to do, does not prevent us profiting by an accumulation of the facts known as to its results, so ought not our ignorance of the whole laws of the hypnotic state to prevent our studying it practically, and applying it beneficially, when we have the power of doing so.[31]

Braid viewed the processes at work as mysterious, powerful and obscure, yet he appreciated their practical use and eventually gained a not-unquestioned acceptance of the validity of their effects from some progressive quarters. It had been a difficult journey. Animal magnetism began as a kind of post-Newtonian force, then became a strange health cure for the rich, before ending up as a something more subtle and harder to define: a means of *influence* that operates upon the mind. It demonstrated to Victorians that the matter of their motivation and agency was complicated. In the decades after Elliotson's astonishing demonstrations with the Okey sisters, hypnotism would become associated with a range of practices, some mysterious and 'occult', and some entirely respectable. Its nature would continue to be a contentious topic in debates about the qualities of 'spirit' and 'mind', into the twentieth century and beyond.

---

[31] James Braid, *Neurypnology*, page 113.

CHAPTER 5

The Electric Miracle

Abstract In this chapter, I examine theories about the nature of electricity, from a mysterious and magical force to a practical one employed to do work. I use Mary Shelley's *Frankenstein* to introduce the range of anxieties and potential opportunities that electricity presented, then explore the arguments about whether or not electricity might be considered a special—or even divine—force. I examine religious interpretations of electrical phenomena, and the utopian optimism regarding its use in communications with the development of the Atlantic Telegraph.

Keywords Electricity • Magnetism • Machine • Atlantic Telegraph • Frankenstein • Mary Shelley • Benjamin Franklin

> The human body is a machine that winds its own springs. It is the living image of perpetual movement (Julien Offray de la Mettrie, *Man a Machine*).[1]

The year 1818 saw the publication of perhaps the greatest work of literature ever created by a teenager: *Frankenstein*. In Mary Shelley's astonishing novel, a natural philosopher sets out to create a human being from an

---

[1] Julien Offray de la Mettrie, trans and notes G C Bussey, *Man a Machine* (Chicago: The Open Court Publishing Co., 1912) page 93.

assemblage of body parts taken illegally from graveyards. When he finally succeeds in giving his creation life, it appals him:

> It was already one in the morning; the rain pattered dismally against the panes, and my candle was nearly burnt out, when by the glimmer of the half-extinguished light, I saw the dull yellow eye of the creature open; it breathed hard, and a convulsive motion agitated its limbs.[2]

The many cinematic dramatizations of this event—most spectacularly presented in James Whale's 1931 version for Universal Pictures—take considerable liberties with the mechanics of the moment. In Whale's interpretation, the body of the creature is elevated to a roof and struck by lightning during a violent electrical storm while Colin Clive's Henry (sic)[3] Frankenstein grows hysterical at the realization of his god-like power. Yet Mary Shelley's original story is much more discreet and mysterious, with the body being brought back to life via some unexplained alchemical process: 'I collected the instruments of life around me, that I might infuse a spark of being into the lifeless thing that lay at my feet'.[4] The apocryphal, kite-flying experiments of Benjamin Franklin have been retro-fitted onto Shelley's narrative to create a truly spectacular visual set-piece. Yet the original narrative gives an accurate picture of the way in which electricity was viewed: with a combination of awe, fear and excitement. By the mid-nineteenth century, even those who could demonstrate electricity in action—known as 'electricians'—still had no idea how this most mysterious of forces was produced. In Britain, the stakes were inordinately high, as electricity became increasingly associated with popular and revolutionary movements. As Iwan Rhys Morus has written:

> If electricity really was the stuff of life as radicals thought, then there was no soul, only organized matter. Without souls there was no need for the Church and without the Church to underpin it, what was left of the English constitution?[5]

---

[2] Mary Shelley, *Frankenstein or The Modern Prometheus* (Oxford: OUP, 1984) page 57.
[3] *Frankenstein*, Dir James Whale, Universal Pictures 1931.
[4] Mary Shelley, ibid.
[5] Iwan Rhys Morus, *Shocking Bodies: Life, Death and Electricity in Victorian England* (Stroud: The History Press, 2011) page 40.

The ancient Greeks had known about electricity. Indeed, its name derives from their word for amber. Thales—one of the revolutionary Ionian thinkers we've already encountered—demonstrated that a static charge was generated when the gemstone was rubbed with fur or hair. Electricity had to be re-interpreted when it was demonstrated to *flow*, and it was probably the eighteenth century's preoccupation with ideas of vital fluids that led to it being referred to as though it were a liquid. Yet for centuries, the power of electricity was used as a magic trick, a means to lift small items or set ablaze a cup of alcohol. It was a baffling but slight mystery, usually deployed for the purposes of entertainment.

One of the earliest demonstrations of electrical power in Britain was a strange combination of experiment and magic act. It took place in 1729 in the alms house at Charterhouse, London. Stephen Gray was born in Canterbury and received only a rudimentary education before he was apprenticed as a fabric dyer. In a trajectory that adumbrated Michael Faraday's, Gray became obsessed with science, particularly astronomy, and this brought him close to the followers of Newton. Despite making a significant contribution to the building of an observatory at Cambridge, illness and poverty led Gray to the Charterhouse, whose building complex also contained a school. As part of a range of practical experiments, Gray built a wooden frame and attached a swing by means of silk cords. He then made perhaps the most dramatic use yet of an extraordinary new invention: the Hauksbee Machine. Francis Hauksbee had been employed by Isaac Newton as his Head of Demonstrations at the Royal Society. In 1705, one of his first presentations was a glass tube on an axle attached to a wheel that spun, generating static electricity. The resulting blue light that flickered in the tube seemed to be the most vibrant demonstration of local electrical force yet.

Stephen Gray's experiment may sound rather alarming and reckless to us: a boy was placed on the swing and connected to the Hauksbee Machine, and with the machine turning and the boy gently swinging, fragments of gold leaf were—apparently miraculously—drawn to his fingertips to the astonishment of the audience. This demonstration formed part of a series of experiments during which Gray displayed not just the existence of electricity but its progression through various conduits. What's particularly noteworthy here is the identity of these early electricians: inquisitive working people whose demonstrations were based on practical observation and experience. In the years before the publication of *Frankenstein*, electricity was increasingly viewed as a potentially liberating new force in a range of ways, and those leading its investigation were certainly not elite scholars based at Oxford and Cambridge. Moreover, these electricians enthusiastically shared information, insight and technology with each other to form a network that was both trans-European and trans-Atlantic.

In Germany, Mathias Bose—the son of a merchant—was involved in a lengthy dialogue with his fellow electricians in Europe. He was Professor of Natural Philosophy at Wittenberg and fascinated by Hauksbee's undervalued invention. Bose built a more powerful version of the machine and—from 1742–1745—used it to create a series of demonstrations that were spectacularly theatrical. In his 'Electric Kiss', a young woman was invited to stand on a block of resin whilst Bose turned the Hauksbee Machine. The woman received a mild static charge. A gentleman from the audience was invited to give her a kiss, suffering a mild shock for his pains. More spectacular was Bose's masterpiece 'The Electric Beatification'. A member of his audience was declared 'king' and invited to sit on a throne. Above them was a metal crown positioned just above the king's head. With the machine being charged offstage and the candles in the room extinguished, the crown of the king spectacularly gave off sparks and the effect of a strange halo. Like Mesmer, Bose's enjoyment of the theatrical elements of demonstration led to him portraying himself as a magician or wizard. Once again, the line between entertainment and educational experiment was blurred, yet the potential power of this extraordinary and occult force was, at last, clear to see.

In Leyden in the Dutch Republic, Pieter van Musschenbroek, Professor of Natural Philosophy and Newtonian, attempted to investigate how electrical charge could be stored. Seeing electricity as an invisible fluid, as his peers did, Musschenbroek set up a glass flask and half-filled it with water. Using the Hauksbee Machine, he attempted to charge the water in the jar via a brass wire but found that the charge didn't remain. According to the professor, an error led to a great discovery. Instead of leaving the glass to sit upon an insulated block, Musschenbroek picked it up whilst turning the Hauksbee Machine. When he stopped turning the machine and put his hand upon the metal lid of the flask, he unwittingly completed a circuit and received a near-fatal electric shock. In January 1746, he shared his findings with a French colleague, describing the power he'd experienced and telling him that he would never dare to repeat the experiment. Alongside the performative elements of the new power, Musschenbroek had also demonstrated its terrifying potential. His electrically charged flask the 'Leyden Jar' proved that electricity could be stored. It was an early version of something future generations would come to rely on heavily: the battery.

By 1746, electrical experimentation had already reached the American colonies. In Benjamin Franklin's autobiography, he recounts a meeting with a Scottish professor called Spence who 'show'd me some electrical experiments' prior to receiving a gift from a member of the Royal Society in London:

> a present of a glass tube, with some account of the use of it in making such experiments. I eagerly seized the opportunity of repeating those I had seen at Boston and, by much practice, acquired great readiness in performing those, also which we had account of from England, adding a number of new ones. I say much practice, for my house was continually full, for some time, with people who came to see these new wonders.[6]

It's interesting that Franklin was frequently supported in his electrical investigations by men working practically with forms of technology derived from Hauksbee, yet his key proposal, that lightning is a form of electricity, was met with opposition from the intellectual elites in both Britain and France. Franklin describes writing to a Dr. Mitchel, a British acquaintance of his, who presented his theory to the Royal Society but 'was laughed at by the connoisseurs'.[7] Franklin's ideas were shared in Britain via a series of privately published pamphlets rather than in volumes of academic discourse produced by universities. In France, Franklin encountered similar problems. Jean-Antoine Nollet, clergyman and 'preceptor in Natural History to the Royal Family', had undertaken his own electrical experiments and refused to believe 'that such a work might come from America'.[8] Nollet's opposition to Franklin's arguments about lightning were later refuted by the French Royal Academy of Science and, in May of 1752, electrician Thomas-Francois Dalibard 'drew lightning from the clouds' close to Paris at Marly-la-Ville by erecting a 40-foot metal rod outside his house during a violent storm. Dalibard used a wine bottle to ground the rod and extracted an electrical charge. For Franklin and other young radicals in Europe and North America, electricity was an example of Enlightenment values practically applied. Anyone could generate

---

[6] Benjamin Franklin, *The Autobiography of Benjamin Franklin* (Philadelphia: Henry Altemus, 1899) page 268.
[7] Ibid., page 269.
[8] Ibid.

electricity using simple and affordable elements. In 1753, a group of young experimenters at the Royal Society awarded Benjamin Franklin a £100 prize in recognition of his work demonstrating 'the sameness of lightning to electricity'.[9]

The principal battleground in the struggle to understand the nature of electricity was Italy. Luigi Galvani was the son of a goldsmith. He became a physician and lecturer in anatomy at the University of Bologna. In 1791, Galvani published *De Viribies Electricatis*, detailing his investigations into what he termed 'animal electricity'. In this account, he described an experiment on the muscles of the legs of a frog he'd prepared for dissection. When the legs were attached to an electrical machine they twitched in response. Galvani also noticed that, when the legs were hung upon brass hooks and attached to an iron fence, a similar reaction took place. As John Gribbin has written: 'he concluded that the convulsions were caused by electricity stored, or manufactured, *in* the muscles of the frog'.[10] For Galvani, this was evidence of a divine spark or vital first principal at work in nature and, since he was working in a university within the territory of the Papal States, this was perhaps the necessary interpretation of the data. Yet this was the period of the French Revolution, and more radical conclusions were possible in cities beyond the Catholic Church's direct influence. At Pavia in Lombardy, in a university controlled by Austria, Alessandro Volta challenged Galvani's conclusions. Volta argued that, far from being contained inside the body, the force twitching the legs of the frog came from outside it. The muscles were merely a conduit, and certainly not a generator or evidence of some innate, holy power. Volta believed that it was the contact between the brass hooks and the iron fence that was the key to the movement Galvani had perceived, a conclusion prompted by Volta's claim that he could 'taste' electricity when he placed coins and a spoon on his tongue. A skilled engineer, Volta made further improvements to the Hauksbee Machine—building on the work of Bose—and then created the masterpiece that would win the debate with Galvani (at least for the moment): the voltaic pile. First described in a letter of 1800 to the Royal Society in London, Volta's invention was a stack of discs of silver and zinc placed one on top of the other and separated by thick paper soaked in salt water. When a wire joined the two sets of discs, a current was generated. In Volta's model, electricity was no longer 'static'.

---

[9] Ibid., page 268.
[10] John Gribbin, *Science: A History 1543–2001* (London: Penguin, 2003) page 289.

It *flowed* consistently—again sounding very much like an invisible fluid. Volta had created a more sophisticated, and safer, variation on the Leyden Jar.

Viewed from the twenty-first century the impact of Volta's discovery is rather blunted. But in Europe, especially in the aftermath of the French Revolution, the concept of portable electrical power was groundbreaking. As we have seen, a framework for practical investigation had been established over the previous decades that enabled working people to participate on equal terms with the elites, and thoughts on the new power and new modes for experimentation could be shared quickly. If Volta was correct, then electricity was a force that could be made to perform work. French electricians—inspired by Franklin—had demonstrated that lightning was simply an intensely powerful manifestation of electricity, so a force usually attributed to God had been harnessed and directed—in small ways admittedly—by ordinary mortals. The full title of Mary Shelley's novel of 1818 is *Frankenstein or The Modern Prometheus*, connecting her story to that of the rebellious Greek god who brought fire to humanity. The significance of Volta's discovery was not lost on the intellectuals of the following decades, who appropriated it in order to justify their own radicalism and defiance of previously sanctified models of social order. Electricity was one component of a sustained ideological challenge to traditional authority.

Galvani died in December 1798, but his theory of animal electricity was still very much alive, despite Volta's publications. In defence of Galvanism—a term even Alessandro Volta used with respect—Galvani's nephew Giovanni Aldini conducted a number of new experiments and documented them in 1803's *An Account of the Late Improvements in Galvanism with a Series of Curious and Interesting Experiments*. His defence of his uncle's theory was partly inspired by recent experiments by the English natural philosopher Henry Cavendish on a fish that could generate an electric charge: the torpedo fish (now known as the electric ray). Aldini passionately asserted that his uncle's position regarding animal electricity had been the correct one but shifted that position to describe a specific and unique fluid generated within the body. He wrote:

> Galvanic fluid is peculiar to the animal machine, independently of the influence of metals, or of any foreign cause…a principle which belongs to the

organization of the human animal...a very energetic fluid, capable of transmission if deriving its origin from the action of animal forces.[11]

For Aldini, the arrangement of organs within the body generated the charge in the same way as the alternating metals of the voltaic pile. For his experiments on the corpse of George Forster 'a malefactor executed at Newgate on Jan 17th 1803'[12] Aldini made use of a decidedly voltaic sequence of 'plates of zinc and as many of copper' presumably in order to stimulate the actions of animal electricity in the deceased body. It's hard to imagine that Mary Shelley hadn't read Aldini's vivid account in her preparatory research for *Frankenstein*. He wrote: 'On the first application of the arcs, the jaw began to quiver, the adjoining muscles were horribly contorted, and the left eye actually opened'.[13] Aldini then attempted to 'excite action in the ventricles' to restart Forster's heart. He didn't succeed. One wonders what the response might have been if he had. Aldini concluded 'That the effects of Galvanism on the human frame differ from those produced by electricity communicated with common electrical machines'.[14] He therefore recommended the continued investigation into animal electricity to potentially 'relieve the sufferings of mankind',[15] the hope being that the vital spark could be rekindled in the recently departed to reinvigorate their own animal electrical fluids.

Events such as the Forster experiment were both scientific demonstration and gruesome spectacle. As Iwan Rhys Morus has observed, there was a conservatism in the process as it emphasized the power of the state and the church over the life and death of the offender. However, in addition to this:

> There was...a dangerous undercurrent to these performances which investigated the boundaries between the living and the non-living. Many Englishmen – Edmund Burke amongst them – listed galvanism amongst the rogue's gallery of malign forces responsible for the French Revolution.[16]

---

[11] John Aldini, *An Account of the Late Improvements in Galvanism with a Series of Curious and Interesting Experiments* (London: Cuthell and Martin, 1803) page 7. Aldini's first name is given as 'John' in English translations of his work.
[12] Ibid., pages 189–203.
[13] Ibid., page 191.
[14] Ibid., page 193.
[15] Ibid., page 203.
[16] Iwan Rhys Morus, page 27.

The protected, special nature of animal electricity allowed for a benign God to remain firmly in the picture, supplying a vital spark that was specific to His creation. Yet the idea of isolating and deploying a power that derived *from* God could be considered radical, defiant and even wicked, and this was deeply troubling to the church-bound intellectual elites of Europe. The argument to establish the nature of electricity ran parallel with debates that pitted religious narratives against new theories of human nature.

Since the unravelling of the Ptolemaic model, humanity had been displaced from the centre of the cosmos and had its planetary home demoted to a revolving supplicant in the court of the sun. The philosophers of the Enlightenment—particularly the French ones—had articulated the idea of Natural Law as a counter to ideas of innate social structure sanctified by religion. Some maintained their faith amidst the passionate exchange of ideas, but some took up positions that were agnostic, and even atheist.

Perhaps the greatest provocation to conventional ideas of faith came from a French physician named Julien Offray de La Mettrie. Like so many disruptive minds of this period, La Mettrie was the son of a merchant. He was also preparing for a career in the church before the febrile intellectual atmosphere of Enlightenment in France led to his turning away from faith in order to study Natural Philosophy. La Mettrie became a medical doctor and—in a rather Newtonian way—used his own body as a source of experimental data. His experience of fever led him to conclude that the process of thought was influenced by forces operating in the nervous system and, primarily, the brain. When his findings were published in 1745, the ensuing critical response was so violent that he was forced to flee to the Dutch Republic, specifically, to the home of the first electrical battery: Leyden. In 1746—two years after Musschenbroek's near-fatal encounter with an electrical current—La Mettrie produced a more-detailed work on his theory of the 'human animal': *L'Homme Machine*, usually translated in English as *Man a Machine*.

In some ways, *Man a Machine* foreshadowed the philosophical works of Kant and Schopenhauer that would be written in the following decades. La Mettrie begins the work with a striking and deliberate rebuttal of his French critics that was strangely reminiscent of ancient Ionian thought: 'Let us conclude boldly that man is a machine, and that in the whole universe there is but *a single substance* differently modified'.[17] He describes his methodology as being empirical with his senses the key mediators of all data. The senses 'induced me to follow reason by lighting the way themselves'.[18] He then slams the door on the idea of metaphysical speculation and occult insight as surely as Kant would in the *Critique of Pure Reason* 39 years later: 'What more do we know of our destiny than of our own origin? Let us then submit to an invincible ignorance on which our happiness depends'.[19] For La Mettrie, the limits of knowledge have been set by our senses and the material and finite nature of our bodies. He concludes: 'He who thinks so will be wise, just, tranquil about his fate, and therefore happy'.[20] It is, in many ways, the most deft and succinct response to the religious faith La Mettrie had once studied.

His philosophy was decidedly Romantic in the sense that La Mettrie viewed the power of nature as all-pervading and transcendent: 'nature has used but one dough and has merely varied the leaven'.[21] Natural Law is 'impossible to destroy…the impress of it on all animals is so strong'.[22] The possibility of God is not entirely removed from the picture, but La Mettrie questions any being that would demand worship from His creations. This was 'a theocretic (sic) truth with very little practical value'.[23] His image of the human is of a wild, noble creature adapting to being:

> thrown by some chance on some spot on the earth's surface, nobody knows how or why, but simply that he must live, and live like the mushrooms which appear from day to day and like those flowers which border the ditches and cover the walls.[24]

Seventy years later in *Frankenstein*, Mary Shelley depicted the creature's first experience of nature in a forest near Ingolstadt:

> I lay by the side of a brook resting from my fatigue, until I felt tormented by hunger and thirst. This roused me from my nearly dormant state, and I ate

---

[17] Julien Offray de la Mettrie, trans and notes G C Bussey, *Man a Machine* (Chicago: The Open Court Publishing Co., 1912) page 148.
[18] Ibid.
[19] Ibid., page 147.
[20] Ibid.
[21] Ibid., page 117.
[22] Ibid., page 118.
[23] Ibid., page 122.
[24] Ibid.

some berries which I found hanging on the trees, or lying on the ground. I slaked my thirst at the brook; and then lying down, was overcome by sleep.[25]

La Mettrie's advice to those who agree with his world view is to embrace a kind of cheerful agnosticism and to view body and soul as consisting of the same material animated by the same force. His theory for the animation of life is strangely alchemical but also familiar: human activity is a combination of processes—a 'chyle' or active fluid is created that enters the blood and 'excites it' creating a 'filtration of spirits which mechanically animate the muscles'. La Mettrie was, of course, describing the process of human digestion and the conversion of food into energy. His conclusion about the nature of life was one of the founding concepts of Enlightenment materialism, based on his own practice as a physician and close observation of his own body. For him, such observation was the only valid starting point for studying the forces at work in nature.

At the same time as the experiments that created the Leyden Jar, in the same city, La Mettrie constructed the idea of a human machine fuelled by organic matter and operating on the principles of perpetual motion. It's unclear as to whether Galvani, Volta or Aldini had read his work, particularly given the outrage it produced outside of progressive and select intellectual circles. Yet we can see the same arguments forming, dissolving and coalescing around a range of concerns from the mid-eighteenth century to the aftermath of the French Revolution: what was the power that animated human life? Was it the same as that which animated other species? Was it produced by the human body itself and the organization of its organs, or was the body a conduit for an occult force of divine origin? La Mettrie died in 1751, almost three decades before the debates of Galvani and Volta were initiated, but one wonders how he would have incorporated ideas of electrical energy and its relationship to life into his argument.

In 1818, the year of *Frankenstein*, the Danish chemist Hans Christian Orsted commenced a series of experiments to show a connection between electricity and magnetism. Two years later, he was ready to reveal his principal discovery: a circular magnetic field generated when current moved through a wire. The effect was demonstrated by the movement of a compass needle. Orsted's discovery led to his being honoured by both the British Royal Society and the French Acadèmie. But rather than offering

---

[25] Mary Shelley, pages 102–103.

an explanation for the existence of electricity, Orsted's work served to make it appear even more mysterious.

Orsted's research was motivated by his interpretation of the writings of Kant, particularly the apparently irreconcilable problem of noumenon and phenomenon we have already considered. As a scientist of faith, Orsted could not believe that a loving God would make the essence of creation undiscoverable to His children. Orsted turned to the writings of the German followers of Nature Philosophy. To Kantians, this denied the limitations of sensory experience that Kant had demarcated and seemed dangerously close to mysticism. But for Orsted, his demonstration of the connection between electricity and magnetism was the first step in a potential, practical unification of all natural forces, and the possible identification of the building blocks of creation itself. The search for a primal force—the key to the strange powers that had been revealed in the century after Newton—now became a principal inspiration for many scientists.

In London, Michael Faraday was one of a number of 'electricians' who had grown up far from the British social elites and their universities. Faraday was the son of a blacksmith whose formal education ended when he was 12. Yet, we might assume, he saw a great deal in the heat of his father's forge, where steam was generated and metals were heated and reshaped, to inspire his perpetually enquiring mind. Apprenticed as a bookbinder, he took every opportunity at the end of the long working day to read copies of the books he had bound. When he could, he attended the lectures of the greatest British chemist of the day: Humphry Davy. Eventually, Faraday was appointed to work at Davy's laboratory. Faraday knew how hard it was for a working-class young man to acquire an education in England in the early nineteenth century. The text of his lecture for young people on *The Chemical History of a Candle*, given at the Royal Institution at Christmas, 1848, when he was 57 years old, displays his personal warmth and desire to spread knowledge to those who were once in a similar position to himself: 'though I stand here with the knowledge of having the words I utter given to the world, yet that shall not deter me from speaking in the same familiar way to those whom I esteem nearest to me on this occasion'.[26] Faraday playfully presented his demonstrations as though they were magic tricks whilst reminding the audience: 'We come here to be philosophers; and I hope you will always remember that

---

[26] Michael Faraday, *The Chemical History of a Candle* (Cromwell Collier, 1962) page 19.

whenever a result happens, especially if it be new, you should say "What is the cause? Why does this occur?"'.[27]

Like Newton, Faraday was driven by a distinctly non-orthodox variation on Christian faith. And, like Orsted, he sought to identify a spiritual element within the material world he was investigating. Had he attended Oxford or Cambridge and become a 'Natural Philosopher', this non-conformism would surely have proved problematic, but Faraday was outside of the mainstream in a few ways. He had been raised a Sandemanian—a sect with distinctly egalitarian tendencies that was strict in its adherence to Bible texts but also emphasized the importance of communal effort. In place of the traditional rite of Communion, Sandemanians prepared a meal to feed their (usually poor) working-class communities. This idea of the scientist as outlier, operating from motives deeper than financial remuneration or patriotism, is important as we consider the progressive and optimistic values that underpinned the process of experimentation in the new world of electro-magnetism.

Building on Orsted's discovery, Faraday converted electric current into continuous movement by moving a magnet in and out of a reel of copper wire. He then developed this model further by turning a copper disc within a magnetic field, displacing electrons that migrated towards the wire. Finally, the electrical current could be made to flow continuously as long as the disc turned. Faraday had created the first motor and the first generator. Electro-magnetism could not only be harnessed, it could be deployed at a distance with the effects of the activity being experienced at the end of the wire. This activity is so familiar to us that it's easy to underestimate the impact of seeing it demonstrated 'live', as Faraday did. To the uninitiated—who numbered all but a handful of practical 'electricians' and some members of the social and intellectual elites—this was either a miraculous event or some kind of magic trick. It would take a long time for the concept of electricity to be domesticated and made safe and, like the steam power of previous generations, it would continue to inspire awe and suspicion for some time to come.

Since Benjamin Franklin had proposed that lightning was really electricity in the previous century, the process of relentless invention had continued in the United States. Samuel Morse was a professional portrait painter from Massachusetts whose clients included the Marquis de Lafayette, the leading French sponsor of the American Revolution. In the

---

[27] Ibid., page 25.

background of Lafayette's full-length portrait, Morse added a small homage to one of his heroes—a bust of Franklin. Morse represented an extraordinary confluence of progressive and conservative ideological and aesthetic forces. He was a republican revolutionary, an aggressive Protestant and an artist working in traditions of Classical epic who became obsessed with the new electro-magnetic science. He also, sadly, endorsed slavery. In 1832, whilst returning to the United States from France by ship, Morse observed the practical demonstrations in electro-magnetism by Charles Jackson of Boston. Morse concluded that the control of an electrical circuit—turned on and off for different durations—could provide a means of relaying messages. Morse put aside his artistic projects to focus on the development of the communication system and, in 1844, sent the first telegraphic message from the Capitol in Washington to the Mount Clare station in Baltimore. It was a quote from the Old Testament *Book of Numbers* (23:23) 'What hath God wrought'. The choice of text is revealing regarding what Morse and his collaborators believed was happening. This new technology was part of a process of exegesis or divine revelation that could lead to peace on earth for all time. As we can see, the Newtonian binding of scientific quest with radical faith was still very much in evidence in the mid- to late nineteenth century. For some, the very mysteries of God were being revealed in strange new ways.

In 1848, Morse had established his system as the primary mode of telegraphic communication, solving the problems of long-distance messaging by adding circuits to boost the range to several miles. Within a few years, the opportunity of Trans-Atlantic communication became a real possibility, backed by the American entrepreneur Cyrus Field. In December 1856, with a range of US and British backers investing a colossal £350,000, the Atlantic Telegraph Company was formed. On 25 July 1858, two ships met in the Atlantic to splice two enormous, insulated cables that were then slowly unwound and paid out onto the ocean floor as the ships returned back to Newfoundland and south-west Ireland, respectively. No single ship could have carried the vast weight of a single-strand cable. Four days after the splice, the cables were attached at their stations and a message of congratulations from Queen Victoria was relayed to President Buchanan. It took 16 hours to arrive, but the event was met with wildly optimistic celebrations on both sides of the ocean.

At the Broadway Church in Boston on 8 August 1858, the Reverend Joseph Copp gave a sermon to commemorate this momentous telegraphic union. His words reflected the euphoria and optimism of the time: 'Fears and misgivings are at an end—the marvellous work is done!'[28] with Copp seeing the cable as an opportunity for a shared communion of faith: 'along the wonderful cable…Christians pray and ministers preach' and, reflecting the earliest message sent by Samuel Morse: 'The Lord Hath Wrought This!'[29] Like the cable itself, Samuel Morse is presented as a conduit for supernatural power:

> Who put it into the heart of the immortal Morse…through years of neglected toil and discouraged hope which have today culminated in the last-greatest wonder of the world…Like Bezaleel, the son of Uri, he was filled with the spirit of God, in wisdom and understanding, and in knowledge…His soul was touched by a spark brighter than electric light…a spark of heavenly wisdom itself.[30]

Morse's miracle does not use the expected divine trajectory by travelling from heaven to earth, but instead moves 'under depths, mysterious, profound and awful, carrying the message of God: "Peace on Earth and good will to men".[31] As we've seen in debates about the nature of 'animal electricity' between Galvani and Volta, there was a clear desire to include faith within the rapidly changing world of technological discovery. The scientist is presented as a kind of new prophet reminiscent of Alexander Pope's epitaph for Isaac Newton intended (but not permitted to be used) for his monument in Westminster Abbey: 'Nature, and Nature's laws, lay hid in Night, God said: "Let Newton be!" and All was Light'.[32] In the *Detailed Report of the Proceedings Had in Commemoration of the Successful Laying of the Atlantic Cable*, published in New York in 1863, a similarly rapturous tone heralds the event: 'Wars and rumours of wars will cease to be!'.[33] The Mayor of New York, Daniel F. Tiemann was quick to celebrate the telegraph as an American achievement whilst maintaining the utopian language of his religious contemporaries:

---

[28] Rev. J A Copp, *The Atlantic Telegraph: As Illustrating the Providence and Benevolent Designs of God. A Discourse, Preached in the Broadway Church, Chelsea, August 8, 1858.* (Boston: T R Marvin and Son, 1858) page 3.
[29] Ibid., page 5.
[30] Ibid., page 6.
[31] Ibid., page 7.
[32] www.westminster-abbey.org/abbey-commemorations/commemorations/sir-isaac-newton
[33] C. T. McClenachan, *Detailed Report of the Proceedings Had in Commemoration of the Successful Laying of the Atlantic Telegraph Cable* (New York: E. Jones, 1863) page ii.

> The genius of Franklin, the patriot and philosopher, lit the way to the brilliant succession of discoveries in electrical science…Morse thus enabling the actions, feelings and sentiments of every people to be communicated with the rapidity of thought.[34]

This idea, of emotional and intellectual discourse across a vast distance, would soon inspire a range of investigations into the nature of consciousness. Many would take startling turns and lead the investigators into mysterious new territories. The celebrations in New York ended with the singing of a specially composed hymn entitled 'For the Laying of the Atlantic Telegraph Cable'. Verse Four is especially interesting:

> Lo! The sunbeam limns our features: Fire and Air we yoke to toil:
> Yea, the lightning from the footstool we have chained in hurtless coil!
> Thou, Oh God, o'er FRANKLIN bending gave to him the electric flame,
> And with clean tongues exultant, MORSE declared thy Holy Name![35]

It's an extraordinarily imaginative blending of biblical ideas with modern technological developments, and it demonstrated a contemporary reluctance to depart from ancient narratives in favour of—say—the radicalism of Mettrie's *Man a Machine*, published 28 years before the American Revolution. Even for those who held Republican and revolutionary values, tested in the aftermath of the War of Independence, the supreme achievement of the cable was surprisingly conservative: to reconnect with the Old World and to receive a message from its Queen. Ominously, a telegraph machine was set up at the commemoration party on Broadway to receive a message from the Lord Mayor of London. It failed to arrive.[36]

The apparently miraculous, divinely sanctioned nature of the Atlantic Telegraph was soon challenged by the disappointing reality of actual events. The cable was slow at best, and messages sent from one side of the ocean were received as gibberish on the other. The idea that electricity was a fine, invisible fluid had been suggested before the invention of the Leyden Jar, and the terms used to describe water were still being routinely applied to electricity a century and a half later. However, the end of the first Atlantic Telegraph also saw the end of the idea of electricity as a fluid. A British electrician—Wildman Whitehouse—assumed that, as with water

---

[34] Ibid., page 2.
[35] Ibid., page 277.
[36] Ibid.

pressure, blockages in the cable might be resolved by strongly increasing the output at the source, in this case the voltage. However, instead of forcing the coded messages through the cable, Whitehouse burned it out, and investors in the United States and Britain lost a fortune. Electricity did not travel like water in a current, but in a way that was far more mysterious. Although electro-magnetism had shown its potential value, scientists were still no closer to understanding its nature.

According to the Irish journalist William Howard Russell, '271 messages were sent from Newfoundland to Valentia (Ireland), and 129 messages from Valentia to Newfoundland in 1858'.[37] In 1865 Russell documented a new attempt to lay a cable under the Atlantic Ocean. In the previous decade, he had already used the telegraph to report the events of the Crimean War, accounts which shocked the public by revealing the appalling conditions in which British troops—especially infantrymen—served. The heroic efforts of Mary Seacole to save the lives of soldiers were first described by Russell, and his ability to gain the trust of regular soldiers and to get them to share their stories led to Lord Raglan, the British commander, forbidding his officers to speak with him. Russell is another example of an outsider who realized the value of the new technology to break the stranglehold of the elites. In his work of 1865, *The Atlantic Telegraph*, Russell offered a guarded optimism regarding the renewed venture somewhat at odds with the wild celebrations of the previous decade:

> Remembering all that has occurred – how well-grounded hopes were deceived, just expectations frustrated – there are still grounds for confidence, absolute as far as the nature of human affairs permits them in any calculation of future events to be, that the year 1866 will witness the consummation of the greatest work of civilized man, and the grandest exposition of the development of the faculties bestowed on him to overcome material difficulties.[38]

Despite further setbacks, the line laid in 1866 worked successfully and two cables functioned profitably until 1869 when the French laid out a competing cable. Ultimately, the cable services were made into a combined company, pooling the interests of three nations by the late 1870s. This added to the utopian optimism regarding international communication

---

[37] William Howard Russell, *The Atlantic Telegraph* (Alpha Editions, 2022) page 115.
[38] Ibid., page 117.

and to the growing belief that technology was to be revered rather than feared. Across Europe and America, the idea of communication between previously intractable distances captured the popular imagination. For some, particularly those Christians who did not conform to the creeds of dominant and national faith groups, it was bound up with concepts of spiritual renewal, and even seen as a sign of the imminent return of Jesus Christ.

Mary Shelley's refusal to describe the exact process of the creature's re-animation in *Frankenstein* allowed her audience to imagine, to play and to engage with the greatest ideas of both science and religion. Her preoccupation with mysterious forces is even signalled in the framing device of the novel, with Captain Walton's letter to his sister declaring there are two potential outcomes to his Arctic mission: to discover a passage 'near the pole' or to ascertain 'the secret of the magnet'.[39] By the end of the eighteenth century, practical displays of electricity were more than the climax of a magic act, or a curious by-product of the interaction of metals. Electricity was a power that could be stored, transported and—most important of all—sustained and directed to do work. By the middle of the nineteenth century, its possibilities extended into global communications. And yet speculation about it was rife and its real nature barely known. Did it 'pervade the universe', like Stukeley's description of Newtonian gravity? Was the human mind a kind of electrical 'field' and, if it was, could it be manipulated, shared or even preserved beyond the body's physical death? Was this electric mind the true iteration of what religious believers termed 'the soul'? These questions led some of the most notable scientists of the time to place their careers and reputations on the line as they searched for answers both inside and outside of the laboratory.

---

[39] Mary Shelley, page 16.

CHAPTER 6

Sensitives, Mediums and Conduits

Abstract In this chapter, I examine Karl von Reichenbach's idea of Od, its connection to Swedenborgian concepts of tremulation, and relationship to theories of electro-magnetism. I discuss ideas of the body as a receptacle and medium for occult power and consider attempts to reconcile mind and body through concepts of gradation of energy. Finally, I examine the contributions of Alfred Russel Wallace to the theory of evolution, mental development and human potential.

Keywords Od • Medium • Sensitive • Evolution • Charles Darwin • Alfred Russel Wallace • Robert Owen • Spiritualism

Now if this source be present in some men and absent in others, it is obvious that, taken from this point of view, there are in effect two classes of men: ordinary men, who have none of all these faculties of sensibility, and those peculiarly subject to excitation, who are excited in the way described on every trifling occasion (Karl von Reichenbach, *Letters on Od and Magnetism*).[1]

---

[1] Karl von Reichenbach, *The Odic Force or Letters on Od and Magnetism*, trans. F D O'Byrne (University Books, 1968) page 13.

Establishing the nature of electricity was an ongoing, global project in the early nineteenth century, intersecting with a range of religious and philosophical discussions. At the centre of these debates was the idea of God's divine power in the world and the uniqueness of creation. Alongside the apparently miraculous manifestations of electrical power, other forces were being hypothesized across the emerging sciences. One of these was termed 'vital force' and its principal advocate in the field of chemistry was Jacob Berzelius. His research would establish a framework for investigation that would involve scholars across Europe and raise questions about the world and its apprehension that would continue far into the nineteenth century.

Berzelius was born in Ostergotland, Sweden, in 1779. He was from a family of church ministers. After the death of his parents, he was apprenticed to a pharmacist before training to become a physician. During this time, he began to experiment with a model of Volta's electric pile, eventually building a version of his own. From this point on, Berzelius became one of the pioneers of electrochemistry, creating a theoretical model known as 'electrochemical dualism'. Although this was superseded soon after, the empirical research performed by Berzelius led to the development of the system of chemical notation.

In 1809, Berzelius put forward the theory that organic compounds could not be synthesized since they contained a unique or 'vital' force. This set the kind of parameter in chemistry that Galvani had proposed in the study of electricity and its action on the body. The forces that ran through living things were unique and separate to other manifestations. However, in 1823 the German chemist Frederick Wohler succeeded in synthesizing urea from an inorganic compound thus displacing the vital force theory. Yet the theory didn't end here and continued through a variety of new iterations.

Wohler was co-editor of a journal of chemistry in Germany, and one of his leading contributors was Karl von Reichenbach. Reichenbach was born in Stuttgart in 1788, the same year as Arthur Schopenhauer. Like Schopenhauer, his education was divided between academic study and practical experience of commerce and industry. Metallurgy was a particular interest, and he established a steelworks in Ternitz, in Lower Austria, and a blast furnace at Gaya, Moravia, along with a range of other assets. According to biographer F D O'Byrne, Reichenbach 'may be said to have held sway like a sort of industrial prince from the Danube to the Rhine'.[2] This success allowed him to devote a large amount of time to his personal scientific research, inspired by Berzelius. Reichenbach named his iteration of vital force 'odic force'.

In his *Letters on Od and Magnetism*, published in 1852, Reichenbach gave an extended explanation of the etymology of the name:

> "Va" in Sanskrit means "to move about." "Vado" in Latin and "vada" in Old Norse means "I go quickly, hurry away, stream forth." Hence "Wodan" in Old Germanic expresses the idea of the "All-transcending"; in the various old idioms it appears as "Wuodan", "Odan" and "Odin", signifying the power penetrating all nature which is ultimately personified as a Germanic deity. "Od" is consequently the word to express a dynamid or force which, with a power that cannot be obstructed, quickly penetrates and courses through everything in the universe.[3]

This odic force had elements in common with existing, mostly German concepts in Nature Philosophy. It was all-embracing and reconciled dualistic ideas of mind and body. Its connection to health and vitality also drew comparisons to mesmerism. But Reichenbach insisted that his idea was anchored far more firmly in empirical investigation. As Richard Noakes has written:

> he was adamant that his work adopted the critical approach to the work of Mesmer and his followers that he believed would placate such formidable critics of mesmerism as the German physiologists Emile du Bois-Reymond and Johannes Muller.[4]

Both scholars—Bois-Reymond had served as assistant in physiology to Muller at the University of Berlin—held formidable reputations and engaged with a wide range of subjects, including 'animal electricity', the development of the nervous system and sense perception. Concepts of vital force were a key interest. Reichenbach's attempts to create a legitimate framework were based on close observation of a diverse range of

---

[2] Karl von Reichenbach, *The Odic Force or Letters on Od and Magnetism*, trans. F D O'Byrne (University Books, 1968) page xvii.
[3] Ibid., page 93.
[4] Richard Noakes, *Physics and Psychics: The Occult and the Sciences in Modern Britain* (Cambridge: CUP, 2019) page 37.

subjects, yet his conclusions often appeared more bizarre than those associated with Franz Anton Mesmer.

Reichenbach appears to have believed that animal magnetism was merely one *phenomenal* aspect of the od. It might be used to restore health, but it was not the force itself. Od, which Reichenbach placed alongside electricity, was omnipresent but could only be detected and directly experienced by a certain class of person, the 'sensitive'. The idea of the sensitive was not entirely unique to Reichenbach, but it would gain in significance through his work and inform future investigations into a number of what would later be termed 'pseudo-scientific' subjects. In brief, on account of some refinement of the nervous system, the sensitive was able to perceive the od in ways others couldn't, and there are some similarities with Swedenborg's theory of 'tremulation'. Their sensitivity was 'the consequence of their hitherto unrecognized peculiarity in the sensory faculty'.[5] Unlike the various forms of 'hysteria' that came under investigation later in the nineteenth century, sensitivity was not gendered, nor did the condition depend on one's social class. In his study of sensitives, Reichenbach encountered a cabinet builder, a weaver and the wife of a hotel keeper. If anything, the newly proposed force seemed stronger in the aspiring working and lower middle classes than in the upper ones.

Swedenborg's concept of tremulation was analogous to musical harmony, but Reichenbach's drew instead upon ideas of polarity that were more akin to Mesmer's post-Newtonian research. Od was like magnetism in that it existed between poles, with the sensitive upset by 'polar likenesses' that caused feelings of unhappiness and even nausea. In the dark, and Reichenbach's investigations frequently required use of a dark room, the sensitive could supposedly 'see' odic energy along with electricity which could be witnessed flowing from a battery. This method of testing would influence attempts to investigate the phenomena associated with spiritualism from the 1860s onwards, with the British chemist William Crookes clearly seeing some connection between the sensitive's apprehension of light and the relatively new practice of photography. For investigators, odic force might be the nature of the soul, a refined form of electro-magnetic field or an entirely new energy source. It's impossible to fully understand the Victorian obsession with telekinesis, telepathy, ghosts and the afterlife without encountering the frequently bizarre theories of Karl von Reichenbach.

---

[5] Karl von Reichenbach, ibid., page 13.

Reichenbach attempted to demonstrate how odic force could explain the appearance of ghosts. On a night in November 1844, he accompanied a female sensitive named Leopoldina Reichel to the cemetery at the town of Grunzing, Austria. The sensitive saw 'fiery phenomena on several of the graves'.[6] Later, in Vienna, the same sensitive had another astonishing encounter as she walked the rows of headstones:

> As she approached, the human-like figures melted away; she recognized the fact that they were no more than luminous clouds, such as she had seen in my dark chamber a thousand times. She now had the courage to go up to them, but only encountered a shining vapour...she was able to whisk it about with the movements of her skirt.[7]

Reichenbach's peer Arthur Schopenhauer explained supernatural manifestations as sudden revelations of the will in nature, but Reichenbach's explanation was just as strange. He explained ghosts—and much other supernatural phenomena—as 'carbonate of ammonium, phosphoretted (sic) hydrogen and other products of putrefaction, known and unknown, which liberate odic light in the process of evaporation. When the putrefaction comes to an end, the lights are quenched – the dead have atoned'.[8] What's extraordinary about the nature of the phenomena he describes is its consistency over time and the sheer number of sensitives who appear to have witnessed the same thing, often together. In London, the same critics who had given mesmerism a hard time were no kinder to the theories of Reichenbach, with the idea of odic force being dismissed and even mocked. As Richard Noakes has written:

> For many medical commentators, the case for od was undermined by the fact that Reichenbach and his chief English language translator and champion, the Scottish academic chemist William Gregory, were physical scientists who seemed to lack the knowledge of physiology and psychology that...was essential to a "right investigation of the phenomena".[9]

Defenders of Reichenbach drew attention to the diverse range of phenomena he investigated and his use of controlled environments to examine test

---

[6] Ibid., page 48.
[7] Ibid.
[8] Ibid., page 49.
[9] Richard Noakes, page 40.

subjects. But it meant little to British sceptics. As with the side-lining of John Elliotson, scholars poured scorn upon the nature of the subjects. The gullible, alleged sensitives had simply seen what Reichenbach wanted them to see. But the concept of a vital force was far from done and would reappear in the scholarship of one of the most highly respected scientists of his time, the co-author of the theory of Natural Selection: Alfred Russel Wallace.

In his work *The Scientific Aspect of the Supernatural*, first published in 1866 and subsequently reprinted in the work *Miracles and Modern Spiritualism*, Wallace placed Reichenbach's work first in a select reading list for those 'desirous of knowing the truth' about the mysterious forces demonstrated by mesmerists and the practitioners of a new faith: spiritualism.[10] The work, and Wallace's sustained interest in the topic, led to conflict with many of his friends and colleagues, including Charles Darwin. But Wallace's belief in such forces was largely consistent with his views regarding evolution and formed part of an all-embracing interest in the origins of human consciousness.

Wallace was another investigator from a humble background who funded much of his own research and placed great value on practical, empirical investigation. He was born in 1823 in Monmouthshire, on the border of Wales and England, the eighth of nine children. His father trained as a lawyer but never practised, choosing a range of business ventures that were largely unsuccessful. Consequently, Wallace rarely knew a settled home life and moved around the country a great deal. His autobiography titled *My Life: A Record of Events and Opinions* published in 1908 describes an upbringing that was culturally rich, with his father knowledgeable about literature and philosophy, but financially insecure. From a young age, Wallace showed a powerful engagement with the natural world, describing the landscapes of Usk and, later, the Hertfordshire countryside in rich detail. His only experience of formal education was a grammar school in Hertford that was distinctly Dickensian in character, with Wallace leaving at the age of 14 to live with one of his brothers in London. Apprenticed to a master builder, Wallace experienced radical politics for the first time:

---

[10] Alfred Russel Wallace, *Miracles and Modern Spiritualism* (London: George Redway, 1896) page 34.

But our evenings were most frequently spent at what was then termed a "Hall of Science" situated in John Street, Tottenham Court Road...It was really a kind of club or mechanics' institute for advanced thinkers and workmen, especially for the followers of Robert Owen, the founder of the Socialist movement in England. Here we sometimes heard lectures on the principles of secularism or agnosticism...It was here that I first made acquaintance with some of Owen's writings, and especially with the wonderful and beneficial work he had carried on for many years at New Lanark.[11]

New Lanark, near the Falls of Clyde in Scotland, was the site of a cotton spinning mill that Owen took over in 1799. He implemented a range of progressive changes in working hours and education that reflected his radical beliefs. Owen's writings had a profound effect on Wallace. He had already rejected his parents' Anglicanism in boyhood and preferred the 'more vigorous and exciting style of preaching' experienced among Dissenter and Quaker communities.[12] But he soon came to turn against religion in total. As he reflected on the idea of 'religious fervour' in 1908: 'as there was no basis of intelligible fact or connected reasoning to satisfy my intellect, this feeling soon left me, and has never returned'.[13] Increasingly, Wallace considered the environment 'which modifies the original character for better or worse' as the determining factor in human development. Knowledge gleaned from another of his brothers—an apprentice surveyor—along with his own voracious reading and research allowed Wallace to obtain work as a teacher in Leicester in 1844. During this time, he formed a friendship with fellow scholar Henry Walter Bates. They shared an interest in entomology and planned an expedition to South America, proposing to fund it by selling insect specimens to collectors. One of their key sources of inspiration was Charles Darwin's *The Voyage of the Beagle*, first published in 1839.

Arriving in Brazil, Wallace and Bates explored sections of the Amazon rainforest, making studies of people and languages as well as of the flora and fauna. It was here that Wallace first considered the effects of territorial differences on the development of species, expanding this study further on an expedition to Asia that was described in a series of academic papers on

---

[11] Alfred Russel Wallace, *My Life: A Record of Events and Opinions* (London: Chapman and Hall, 1905) page 86.

[12] Ibid., page 78.

[13] Ibid.

the Malay Archipelago in 1854.[14] Wallace had long been convinced of the truth of evolution but struggled to understand its mechanism. During a bout of fever in January 1858, he claimed to have seen a solution in his mind:

> Why do some die and some live? And the answer was clearly that, on the whole the best fitted live. From the effects of disease the most healthy escaped; from enemies the strongest, the swiftest or the most cunning; from famine, the best hunters or those with the best digestion and so on. Then it suddenly flashed upon me that this self-acting process would necessarily improve the race because in every generation the inferior would inevitably be killed off and the superior would remain-that is, the fittest would survive.[15]

His vast collection of specimens, acquired over years of exploration, evidenced the emergence of new traits and physical changes over time. Wallace described his theory in a paper titled 'On the Tendency of Varieties to Depart Indefinitely from the Original Type'. Having corresponded with Darwin previously, Wallace wrote to him again. In his autobiography, he described the older naturalist's response: 'The effect of my paper on Darwin was at first almost paralyzing. He had, as I afterwards learnt, hit upon the same idea as my own twenty years earlier'.[16] Although there were differences between the theories of Wallace and Darwin, Darwin realized that both had identified the missing element in the origin of species as 'Natural Selection'. Wallace generously deferred to Darwin as the first author on the subject, and both of their papers were presented at the Linnean Society on 1 July 1858. Initially, the theory was known as 'Wallace-Darwinism', but Wallace's gentlemanly acceptance of Darwin's earlier discovery has led to the latter's name being the one exclusively associated with the theory. Wallace, from a humbler background and working constantly in the field, deserves a far greater measure of credit than he presently receives.

In 1867, nine years after the publication of the joint papers on Natural Selection, Wallace joined one of the committees of the London Dialectical Society. The society was made up of professionals from a range of different backgrounds who engaged in investigating and debating contemporary philosophical questions, including the claims of so-called mediums. Thus

---

[14] Published as a single volume: *The Malay Archipelago* in 1869.
[15] Ibid., page 362.
[16] Ibid., page 403.

developed a connection between Wallace and the emerging faith of spiritualism.[17] This interest strained Wallace's friendship with Darwin and created cracks in his relationship with the wider scientific community. As we will also see in the cases of William Crookes and Oliver Lodge, the connection with the 'supernatural' could stall careers in the Victorian period. To his credit, Wallace remained open-minded to all phenomena and believed, somewhat optimistically, that others would approach investigation with the same lack of prejudice. The committees of lay investigators from the professions probably represented a more comfortable social world for him than the university-based elites, but Wallace was an optimist who deferred, perhaps too much, to those scientists in the social classes above him. As Ross A. Slotten has written of him in *The Life of Alfred Russel Wallace: The Heretic in Darwin's Court*:

> Wallace felt certain that if any of his scientific friends could witness what he had seen in his own house "under test conditions," they would be satisfied that the phenomena were genuine, even if they disagreed with the explanation suggested by spiritualists.[18]

However, Wallace struggled to get colleagues to attend séances and, even if they did, they rarely responded positively. In his book *Miracles and Modern Spiritualism*, first published in 1874, Wallace conducted a defence of Reichenbach's odic force against those scholars who simply dismissed it out of hand without attending demonstrations. He began by attacking the work of the previous century's greatest sceptic and empiricist, David Hume. Hume's work *Of Miracles* was part of his 1748 work *An Enquiry Concerning Human Understanding* but it wasn't included in a published version until after the philosopher's death in 1766. This was due to fears of a potentially violent response from some members of the congregation in Scotland. In the section, Hume defines a miracle as 'a transgression of a law of nature by a particular volition of the Deity, or by the interposition of some invisible agent'.[19] Hume states that the preponderance of evidence against a miracle will always weigh heavier since the miraculous

---

[17] Wallace had already begun to attend séances in 1866 at the invitation of his sister, Fanny Sims.

[18] Ross A. Slotten, *The Heretic in Darwin's Court* (New York: Columbia University Press, 2004) page 244.

[19] David Hume, *An Enquiry Concerning Human Understanding* (Chicago: The Open Court Publishing Co., 1900) page 121.

event defies the regularity of everyday experience. Wallace rejected this argument since it assumes that all the laws of nature are known, and no new ones are to be discovered. He cited electricity and its journey from mystical, occult force to its practical use in everyday life as an example. He also asserted that the credibility of the witnesses is key:

> no unprejudiced individual can fail to acknowledge that the researches of Reichenbach have established the existence of a vast and connected series of new and important natural phenomena. Doctors Gregory and Ashburner in England state that they have repeated several of Reichenbach's experiments under test conditions and have found them quite accurate.[20]

There appear to be two versions of Wallace at work in his scholarship: the practical naturalist relentlessly documenting his theory with practical examples, and the more credulous believer in phenomena we might ascribe to imagination, suggestion or trickery on the part of the subject. Wallace attended many séances conducted by those who were later exposed in what Victorians termed 'imposture' or fraud. The truth of Wallace's position resides in his theory of mind. Whereas Darwin concluded that human development could be entirely understood by the mechanism of Natural Selection, Wallace believed that consciousness was so unique that it fell outside of the laws he himself had described. As he wrote of odic force in *Miracles and Modern Spiritualism*:

> the silence or contempt of our modern scientific men cannot blind the world any longer to those grand and mysterious phenomena of mind, the investigation of which can alone conduct us to a knowledge of what we really are.[21]

Wallace concluded that 'Spirit is mind; the brain and nerves are but the magnetic battery and telegraph by means of which spirit communicates with the outer world'.[22] In his theory, certain 'sensitives' (or 'mediums') were better adapted to detecting the invisible forces working around us due to a refinement of their nerves and heightening of their senses. As with Reichenbach, Wallace saw 'sensitives' not simply among the social elites, but in a range of different men and women. Rather than being

---

[20] Alfred Russel Wallace, *Miracles and Modern Spiritualism*, page 57.
[21] Ibid., page 64.
[22] Ibid., page 108.

reactionary, Wallace's belief in miracles was part of the advance of human development. It was:

> progress through a more advanced state of existence, a view which should commend itself to men of science as being in itself probable, and in striking contrast with the doctrines of theologians, which place a wide gulf between the mental and more (sic) nature of man in his present and in his future state of existence.[23]

Wallace viewed the mind as falling outside of the laws of a materialistic version of evolution as proposed by Charles Darwin, but he also argued that the mind or spirit was still engaged in a process of unfolding. The higher levels of perception might be cultivated by a more civilized mode of living that reduced the emphasis on physical labour, as Robert Owen had proposed in his management of the mill at New Lanark, and encouraged the cultivation of mental power and the attributes of the sensitive. As I've already suggested, it has always been tempting for the investigator to place a human face upon mysterious forces and to seek to show their action is bound up with the journey of humanity. I believe Wallace was doing that here, attempting to assert a progressive teleology that gave meaning to life. Even Reichenbach's near-exact contemporary Schopenhauer, a believer in a singular, blind and uncaring force of will that underpinned the universe, elevated the importance of art in the amelioration of the will's many cruelties. Human values could not be entirely subsumed by nature's wild forces.

To his critics, Wallace's views represented a backwards step—he appeared to be suggesting the brain was in a special, separate category to the body's other organs. However, his focus of interest was not simply the materiality of the brain, but the concept of *consciousness*. Wallace viewed the mind as separate from the physical brain in the sense that it might be a kind of field that is constantly interacting with the world to generate reality. In this way, it could be argued, Wallace was another natural philosopher attempting to resolve the absolute binary of noumenon and phenomenon that was proposed in the previous century by Kant. Certain individuals or 'sensitives' might be receptive to different bandwidths of experience. Just as dogs can hear higher frequencies of sound, so some humans might be able to experience beyond the field of perception defined

---

[23] Ibid., page 109.

as regular or habitual. Perhaps these higher versions of the senses could be cultivated, particularly if people worked in more humane environments, such as those proposed in an Owenite utopia. Could humans one day experience the noumenal reality? Given Wallace's knowledge of the journey of species and their traits over time, this is not as fanciful as it sounds.

It's possible that Wallace's reputation has suffered over time due to his endorsement of the work of Reichenbach and defence of spiritualism. However, I have argued that these interests were an integral part of Wallace's approach to science: he ruled nothing out, and since he had not grown up among the university-based elites, he felt compelled to investigate a range of phenomena that reached beyond the purely theoretical. Like Reichenbach, Wallace saw manifestations of od, spirit and vital force in a diverse range of people. As a boy raised in genteel poverty on the Welsh border, an apprentice in London's building trades, and then an explorer in South America and Asia, he had encountered a far more diverse group of humans in terms of class and ethnicity than many of his university-educated peers. He was deeply receptive to ideas from a range of sources. He was undeniably credulous in his support of some of the mediums I'll discuss shortly, and he certainly underestimated the power of suggestion and the mind's capacity to be duped by highly capable frauds. Yet there is a gentleness, determination, and generosity in the scholarship of Alfred Russel Wallace that sits at odds (or 'ods', perhaps?) with the often high-handed and dismissive investigations conducted by organizations such as the Cambridge-based Society for Psychical Research in the later Victorian period. The problem of consciousness, of the relationship between the perceiver and perceived, is very much current, and Wallace's attempts to uncover the mystery ought to have been viewed far more favourably.

CHAPTER 7

The Brief Life of Psychic Force

Abstract In this chapter, I explore the influence of spiritualism upon scientific enquiry in the Victorian period. Spiritualist practice incorporated some of the elements of Mesmer's work, including the idea of séance. Focusing on the career of the medium Daniel Dunglas Home, I trace the widespread interest in spiritualist phenomena and the attempts to put it to the test by the chemist William Crookes and others. These experiments divided the scientific community, as well as adding to the increasingly complex debates upon the nature of the human mind.

Keywords Séance • Spiritualism • Medium • Psychic force • Daniel Dunglas Home • William Crookes • Telepathy • Telekinesis

It has been attested that metaphysical speculation is a thing of the past, and that physical science has extirpated it. The discussion of the categories of existence, however, does not appear to be in danger of coming to an end in our time, and the exercise of speculation continues as fascinating to every fresh mind as it was in the days of Thales (James Clerk Maxwell, *British Association Report*).[1]

---

[1] British Association Report, Vol. XL, September 1870.

In 1865, while William Russell was writing with guarded optimism about the prospects of the new iteration of the Atlantic Telegraph, and Alfred Russel Wallace was pondering if the mind could be considered separate from the brain, the Professor of Natural Philosophy at King's College London, James Clerk Maxwell, was completing a groundbreaking new work: *A Dynamical Theory of the Electro-Magnetic Field*. Maxwell had been experimenting with magnets since his student days in Edinburgh and continued in the tradition of practical experimentation and demonstration established by Faraday. In a work that was a key influence on the young Albert Einstein,[2] Maxwell ultimately demonstrated that electric and magnetic fields travel in the form of waves and that light itself is a kind of electro-magnetism. This was a crucial discovery in the development of modern physics. A decade earlier, Maxwell had solved the problem of Saturn's rings—what were they composed of and why didn't they fly off or crash into the planet?—by concluding the phenomena were composed of tiny particles in independent orbits.[3] By 1870, he was writing about the vastness of the range of forces at work in the universe, but also revealing their intimate connection to our earthly existence:

> the spectroscopic examination of the light of the sun and stars shows that, in regions the distance of which we can only feebly imagine, there are molecules vibrating in as exact unison with the molecules of terrestrial hydrogen as two tuning-forks tuned to concert pitch, or two watches regulated to solar time.[4]

Just as Newton had demonstrated the universal and yet familiar power of gravity, so Maxwell described mysterious powers at work within the very substance of the stars and planets that were strangely close to hand. And, as with Newton, multiple and creative variations of Maxwell's theories would be applied to explain a broad range of strange phenomena. Some were credible, some were based on misunderstandings and some were wildly speculative.

---

[2] Walter Isaacson, *Einstein: His Life and Universe* (London: Simon and Schuster, 2007) pages 91–92.

[3] On 12th November 1980, the Voyager One space probe proved Maxwell's mathematically derived solution to be correct.

[4] James Clerk Maxwell, ed. P. Harman, *The Scientific Letters and Papers of James Clerk Maxwell, Vol. II* (Cambridge: CUP, 1995) page 588.

Like many of his peers, Maxwell was inspired by the French work *Le Mecanique Celeste* by Pierre Simon Laplace. This work contained the 'nebular hypothesis', the idea that the planets had formed from a cloud of debris around the sun, a speculation also made by Kant in 1755 that is now known as the Kant-Laplace hypothesis. This led to the astonishing possibility that the composition of the earth and everything upon it might derive from the outcome of a single, celestial event. Although Maxwell eventually found himself in opposition to the mechanics of the Kant-Laplace hypothesis, his own work certainly supported the concept of a universe of shifting matter in perpetual motion. This raised a range of tantalizing questions about human origins, the connectedness of mind and matter and the composition of space itself. These questions would come to preoccupy a number of very different individuals in the second half of the nineteenth century.

It's important to point out here that the scientists and philosophers of this time were operating in a period when scientific categories were still forming. Consequently, there was a sense of openness in their discourse and a willingness to consider a range of experiences and some very different sorts of data. Cromwell Varley, an engineer and one of the principle British consultants on the development of the Atlantic Telegraph, became deeply interested in the concept of the mind as an electrical system that could possibly transfer information beyond the confines of the brain. Like Elliotson, Varley was convinced that the discoveries of Franz Anton Mesmer were not entirely without merit, and that there was a kind of potential, mental power that might be used to relieve sickness. Despite the victory of Volta over Galvani in the matter of 'animal electricity' several decades before, the possibility of invisible human power generated in the body was a still a matter of great interest. On 25 May 1869, Varley gave evidence to the Committees of the London Dialectical Society—to which Alfred Russel Wallace was also a contributor—regarding a strange experience he and his wife had shared:

> (Mrs Varley) was subject to nervous headaches…She was only temporarily relieved (by mesmerism) and one day, while she was entranced on the couch, I was thinking whether I could permanently cure her. She answered my thought. I considered this very strange and I asked her – still mentally – whether she was answering my thought; she replied "Yes." I then asked if there were any means by which a permanent cure could be effected. She replied "Yes; if you bring on the fit out of its proper course you will disturb

its harmony, and I shall be cured." I did so – by the exercise of will – and by bringing on the fits at intermediate periods, she was cured permanently.[5]

Whatever the actual nature of this experience, Varley was taking a considerable risk in talking about it publicly. Like Michael Faraday, he was raised in the Scottish Sandemanian movement, although Varley had come to reject the faith. He was also from a family that claimed descent from the puritan Oliver Cromwell. His talent for engineering had seen him become the senior adviser to the Electric Telegraph Company by the age of 20 and, by 1861 he effectively ran the organization. In his evidence to the Committees, he stated that he had first encountered strange 'psychic' energies in the 1850s and believed them to be the result of 'electrical force' but concluded finally that this was not a viable hypothesis. This energy was akin to electro-magnetism in some ways but produced differently and in greater abundance in certain individuals. This brought his views closer to the idea of the 'sensitive' as described in the research of Reichenbach. But Varley believed the phenomena might be evidence of a 'human telegraph' that allowed the mind to project its contents, perhaps as some kind of wave or field. Varley's passionately held interest led to the involvement of one of the most distinguished chemists of the period: William Crookes.

Crookes was the son of a tailor. In 1848, aged 16, he enrolled at the Royal College of Chemistry where he funded his own study, including his laboratory apparatus, by becoming a practical demonstrator. This was a tradition that went back, of course, to Francis Hauksbee at the Royal Society under Newton. In 1859, Crookes founded the *Chemical News* periodical and was a regular contributor to the *Quarterly Journal of Science*. Even by the standards of the Victorian scientist, Crookes was prolific and profoundly engaged with developments in the world around him. His principal interest was in the new field of spectroscopy which led to the discovery for which he is perhaps best known: the element thallium. With the support of Varley, Crookes' investigation of the conductivity of electricity through gases led to the discovery of cathode rays and, in 1863, Crookes became a member of the Royal Society. His principal research interest was the nature of matter and the various forms in which it could

---

[5] *Report on Spiritualism of the Committee of the London Dialectical Society Together with the Evidence Oral and Written and a Selection from the Correspondence* (London: Longman, Green, Reader and Dyer, 1871) pages 157–158.

exist. In the spring of 1870, Crookes announced his intention to investigate the strange forces described by Varley and Russel Wallace and discussed by the London Dialectical Society under strict, laboratory conditions. He wrote in the *Quarterly Journal of Science*:

> The spiritualist tells of heavy articles of furniture moving...but the man of science is justified in doubting the accuracy of the former observations if the same force is powerless to move the index of his instrument one poor degree.[6]

William Crookes' subject for this rigorous series of experiments was a medium called Daniel Dunglas Home. Home had come to Britain from the United States in the wake of the women who were said to have begun the spiritualist 'awakening': Kate and Margaret Fox. The new 'science' of spiritualism requires a little explanation and contextualization. In the spring of 1848, in rural upstate New York, the Fox Sisters experienced a number of bizarre events in the family home, including loud knocking, the displacement of objects and movement of furniture, supposedly by some invisible agency. The girls, from a devoutly Methodist background, claimed they were in contact with the restive spirit of a dead peddler. The extraordinary occurrences attracted the interest of both the religious and academic communities. The girls, joined by their older sister Leah Fox Fish, took to the stage to demonstrate their alleged ability to act as conduits for some decidedly strange forces in the form of loud knocks or 'spirit raps' from the next world (Varley later described these as 'a chorus... 50 hammers all striking rapidly could hardly reproduce').[7] Their demonstrations, which took place in the nearby town of Rochester, attracted large audiences and Kate and Margaret, aged 12 and 14, respectively, became celebrities, with Leah working effectively as an impresario, promoting her sisters and planning tours of their séances as far afield as the Midwest. The sisters' new-found powers led them to refashion themselves over time into the evangelists of a new religion that claimed to offer practical evidence of a new form of mental power, and even evidence of the human survival of death: spiritualism. Spiritualist demonstrators offered audiences a rich yet unsettling mix of demonstration and theatrical spectacle. Belief in the phenomena of the Fox sisters was widespread and

---

[6]William Crookes, *Researches in the Phenomena of Spiritualism Reprinted from the Quarterly Journal of Science*, (London: J Burns, 1874) page 6.
[7] *Report on Spiritualism*, page 165.

viewed by some as evidence of divine exegesis or some new phase of human development. The resulting movement was, perhaps, an expression of a deeply felt dissatisfaction with conventional faith and its doctrines, but also embodied a sense of optimism about human potential and the possibility of social change. The new religion allowed the comfort of contact with lost family members and unveiled a next world or 'Summerland' that was both practical and idealistic, more like an Owenite utopia than a traditional image of heaven. From a scientific viewpoint, spiritualism offered the possibility of demonstrable and measurable proof of faith, both now and in the hereafter.

Four years after the experiences of the Fox sisters in Hydesville, Daniel Dunglas Home laid claim to similar powers. Most significantly, he offered to demonstrate them to observers under controlled conditions. Home was born in Currie, near Edinburgh in 1833. He emigrated to the United States with his aunt, Mary Cook, at some point between 1838 and 1841. The facts of his life and career are given by the British-born American medium Emma Britten, the spiritualist press and a number of admirers, including Wyndham Quinn, Lord Adare. Quinn published *Experiences in Spiritualism with Mr DD Home* in 1869. Home was able to demonstrate what was termed 'a spiritually magnetic state' during which he passed on messages supposedly from spirits to their living relatives. Home became a sensation among the Spiritualist community, and he sailed to England in 1855 at the invitation of British admirers. He took rooms in Jermyn Street, London, and proceeded to give séances. In the summer of 1855, one of his sitters included the playwright Edward Bulwer Lytton who seized a spirit hand and was 'half dragged under the table'.[8] Although refusing Home a public endorsement, Lytton remained convinced that Home had demonstrated some as-yet unknown form of natural force.[9] In the same year, Home conducted a séance for Elizabeth Barrett Browning who was, as Adam Roberts describes 'energetically enthusiastic about séances, table-rappings, hauntings and other aspects of the Victorian supernatural'.[10] She became a believer in Home, to the apparent consternation of her husband, Robert Browning, who composed a vitriolic poem

---

[8] Peter Lamont, *The First Psychic: The Peculiar Mystery of a Notorious Victorian Wizard* (London: Abacus, 2006) page 48.

[9] Ibid., page 49.

[10] Adam Roberts, 'Browning, reincarnation and the resuscitation of the dead' in *The Victorian Supernatural*, eds. Bown, Burdett and Thurschwell (Cambridge: CUP, 2004) page 109.

called *Mr Sludge The "Medium"* in 1859, in which a medium confesses that his work is a series of fraudulent illusions and inept attempts at mesmerism. Browning was so concerned the poem would upset his wife that he withheld it from publication until after her death in 1864.

Home did not charge for his séances, as many others did, and did not perform what might be termed an act or routine. He was supported by wealthy spiritualists, who made gifts of money and jewellery to him in return for private séances. Often, these sessions occurred without the sitters experiencing any phenomena at all. In 1869, Home was interviewed by committee members of the London Dialectical Society. Asked to describe a trance, he replied:

> In a trance I see spirits connected with persons present. Those spirits take possession of me; my voice is like theirs. I have a particularly mobile face, as you may see, and sometimes take a sort of identity with the spirits who are in communication through me…I have never been mesmerised, and cannot mesmerise. I have an exceedingly soothing power, and exceedingly gentle way of approaching anyone, whether well or ill, and they like to have me near them.[11]

Home united aspects of the work of Mesmer, the Marquis de Puyser, Dupotet and the Fox sisters to create a new kind of supernatural demonstration that was far too performative for the tastes of many scientific observers, but was frequently breathtaking to others.

Varley and Crookes probably met frequently through their various professional and gentlemanly associations. Additionally, Crookes' brother Philip had worked on a proposed telegraph cable project to join Florida and Cuba. In September 1867, William received the news that Philip had died of typhoid during the exploratory phase. He sued the company for negligence, losing the case and ending up in considerable debt. It's possible that the loss of his brother drew Crookes closer to Varley and his circle of 'scientific spiritualists'. By 1871, Crookes had decided on an experimental framework for putting Home's powers to the test, and the medium had agreed to all the requested controls. The results of these extraordinary investigations would be reported in the *Quarterly Journal of Science* and subsequently published under the title *Experimental Investigation on Psychic Force*.

---

[11] *Report on Spiritualism*, page 189.

Crookes invited Daniel Home to his laboratory at Morning Road, where the chemist had meticulously set up his experiments. To ensure Home could not conceal any magical or theatrical machinery, Crookes first visited the medium at his apartments, as he described in his account in the *Quarterly Journal of Science* of July, 1871:

> in the afternoon I called for Mr. Home at his apartments, and when there he suggested that, as he had to change his dress, perhaps I should not object to continue our conversation in the bedroom. I am, therefore, enabled to state positively that no machinery, apparatus or contrivance of any sort was secreted about his person.[12]

At the laboratory, Crookes introduced Home to the two observers he'd invited to witness the experiments, men with impeccable scientific and social credentials. The first was William Huggins, an astronomer who had built his own observatory at Tulse Hill in south London, presumably recruited because of his interests in photography and spectroscopy. In 1867, Huggins had won the Royal Society's Gold Medal for research conducted with his wife, Margaret Lindsay, and the chemist William Allen on the nature of astronomical phenomena, leading to the differentiation between galaxies and nebulae. The second was Edward Cox, a lawyer and former MP with a long-standing interest in accounts of mesmerism and its effects.[13] As Crookes later pointed out, he invited a number of leading scientists in his London circle to attend but met with little success. It's probably reasonable to assume that these refusals revealed a desire to avoid interaction with phenomena that were generally associated with magicians and the wilder elements of the spiritualist movement. This mirrored the experience of Alfred Russel Wallace, as we've already seen.

Crookes had shown through his editorship of *Chemical News* that he was keen to speak to all, and to investigate many types of phenomena. Indeed, he believed it was the purpose of science to do so. The fact that the London Dialectical Society had published a series of interviews with a range of different observers demonstrated to William Crookes that this was a matter of concern to the public and that his duty as a scientist was to investigate it. In this regard, Crookes operated beyond the respectable

---

[12] William Crookes, page 11.

[13] Varley, it appears, was not in attendance although he would be involved in subsequent experiments.

constraints of the Victorian social elites and showed courage in attempting to examine a controversial figure who had been so comprehensively attacked in verse by Robert Browning. Crookes was his own man, his own scientist, and whatever the nature of his conclusions, he acted from powerful convictions and employed a consistent and coherent methodology, certainly for these early investigations into mediumship.

With Home's clothing checked prior to his arrival at the Mornington Road laboratory, Crookes, Huggins and Cox could begin the first experiment 'in a large room lighted by gas'[14] with Charles Gimingham, Crookes' lab assistant, also present. Gimingham had been trained by Crookes and had worked as a demonstrator for some time. One of Home's most remarkable feats was his ability to make musical instruments play without any direct contact, and Crookes set up an appropriate test using an accordion:

> The apparatus prepared for the purpose of testing the movements of the accordion, consisted of a cage, formed of two wooden hoops, respectively one foot ten inches and two feet diameter, connected together by twelve narrow laths, each one foot ten inches long, as to form a drum-shaped frame, open at the top and bottom; round this 50 yards of insulated copper wire were wound in twenty-four rounds, each being rather less than an inch from its neighbour. These horizontal strands of wire were then netted together firmly with string, so as to form meshes rather less than two inches long by one inch high. The height of this cage was such that it would just slip under my dining table, but be too close to the top to allow of the hand being introduced into the interior, or to admit of a foot being pushed underneath it. In another room were two Grove's cells, wires being led from them into the dining room for connection, if desirable, with the wire surrounding the cage. The accordion was a new one, having been purchased by myself for the purpose of these experiments at Wheatstone's in Conduit Street. Mr Home had neither handled nor seen the instrument before the commencement of the test experiments…Mr Home sat in a low easy chair at the side of the table. In front of him under the table was the aforesaid cage, one of his legs being on each side of it. I sat close to him on his left, and another observer sat close to him on his right, the rest of the party being seated at convenient distances around the table.[15]

---

[14] Ibid., page 10.
[15] Ibid., page 11.

What took place next has been the matter of much discussion and debate. Home took hold of the accordion 'between the thumb and middle finger of one hand at the opposite end to the keys'.[16] With the lab assistant taking up a position under the table, and Home flanked by two of the observers, the accordion 'was seen…to be waving about in a curious manner; then sounds came from it, and finally several notes were played in succession'. From his vantage point on the floor, the lab assistant reported 'the accordion was expanding and contracting'. After some moments 'a simple air was played'[17] and, to the astonishment of all present 'Mr Home then removed his hand altogether from the accordion, taking it quite out of the cage, and placed it in the hand of the person next to him. The instrument then continued to play, no person touching it and no hand being near it'. Crookes then employed the battery to run a mild current through the insulated wire of the cage—presumably to see if some magnetic element was present—and the accordion 'sounded and moved about vigorously'. Crookes wrote that it was 'impossible to say' if the current had impacted on the behaviour seen within the cage. In further tests with the accordion it was observed 'distinctly floating about' and played itself again. Home, Crookes observed, was 'not moving a muscle'.[18]

In a second test, a balance was set up comprising of a mahogany board '36 inches long by nine and a half inches wide and one inch thick':

> One end of the board rested on a firm table, whilst the other end was supported by a spring balance hanging from a substantial tripod stand. The balance was fitted with a self-registering index, in such a manner that it would record the maximum weight indicated by the pointer. The apparatus was adjusted so that the mahogany board was horizontal, its foot resting flat on the support. In this position its weight was three pounds as marked by the pointer of the balance.[19]

Home had been kept apart from the balance apparatus. He sat before the board and placed his fingers 'lightly on the extreme end'.[20] Again, flanked by two of the observers and exerting no obvious pressure:

---

[16] Ibid., page 12.
[17] Ibid., page 13.
[18] Ibid., page 14.
[19] Ibid., page 15.
[20] Ibid.

the end of the board was seen to oscillate slowly up and down…On looking immediately afterwards at the automatic register, we saw that the index had at one time descended as low as nine pounds, showing a maximum pull of six pounds upon a board whose normal weight was three pounds…I scarcely need add that his hands were closely guarded by all in the room.[21]

When the results of the experiments with Home were published, Crookes faced a serious backlash from the scientific community, many of whom he'd challenged to attend the investigation. In his defence in the *Quarterly Journal*, he drew attention to the 'carefully arranged apparatus and…the presence of irreproachable witnesses'[22] at the Mornington Road laboratory. Crookes was convinced that he had observed a demonstration of some as yet undiscovered force in nature that within its conduits is 'power very variable and at times entirely absent'.[23] He restated his adherence to scientific method and advised the investigating 'philosopher' that: 'Romantic and superstitious ideas should be entirely banished, and the steps of his investigation should be guided by intellect as cold and passionless as the instrument he uses'.[24] In a subsequent letter dated 8 June 1871, Edward Cox attempted to give the force he'd witnessed an official name:

> I venture to suggest that the force be termed the Psychic Force; the persons in whom it is manifested in extraordinary power Psychics, and the science relating to it Psychism, as being a branch of Psychology. Permit me also to propose the early formation of a Psychological Society, purposely for the promotion by means of experiment, papers and discussion of the study of that hitherto neglected science.[25]

In 1875, Cox established just such an investigative body, inspired by the experiments he'd witnessed with Crookes and Huggins. Cox sought to disconnect the new organization from the idea of ghosts and other such 'romantic' manifestations. Just as Elliotson had focused on the therapeutic and practical elements of mesmerism—the trance that permitted painless surgery—the Psychological Society would focus upon manifestations of mental power, what would soon be termed 'telepathy' and 'telekinesis'.

---

[21] Ibid.
[22] Ibid., page 17.
[23] Ibid.
[24] Ibid.
[25] Ibid., page 19.

Cromwell Varley had concluded that the force demonstrated by Daniel Dunglas Home was electrical in nature. With Maxwell's bringing together of light, electricity and magnetism as manifestations of the same essential force, operating as fields and waves, Crookes now had the opportunity to integrate his new findings into a much bigger picture. He was very well acquainted with the work of Reichenbach, whose subjects frequently 'saw' forces as illuminations whilst simultaneously experiencing their physical impact. As we've seen, Reichenbach's accounts of his experiments were amongst those most highly valued by Alfred Russel Wallace. Crooke's next step was to investigate how psychic force travelled from body to body and from mind to mind. But this investigation would risk his credibility, as the boundary between 'legitimate' investigation and speculative or 'occult' enquiry became increasingly concrete and impermeable.

CHAPTER 8

A Psychic Body

Abstract In this chapter, I examine William Crookes' further investigations into the nature of spiritualist phenomena, focusing on the career of the medium Florence Cook. Cook appeared to demonstrate some extraordinary powers in the séance room. The documentation of these powers led to some scientists questioning Crookes' abilities as an examiner, and to a dispute between Crookes and the sceptic W B Carpenter. This dispute fundamentally called into question the legitimacy of enquiries into mysterious and occult forces.

Keywords Medium • Ghost • Spiritualism • Psychic • Séance • Florence Cook • Katie King • William Crookes • W B Carpenter • London Dialectical Society

> Let those who are inclined to judge Miss Cook harshly suspend their judgement until I bring forward positive evidence which I think will be sufficient to settle the question (William Crookes, *The Quarterly Journal of Science*, 1874).[1]

---

[1] William Crookes, page 104.

The experiments described in the previous chapter—witnessed by two other eminent Victorians and an experienced and well-trained laboratory assistant—revealed one of two things: either Daniel Dunglas Home had demonstrated telekinesis in a room, or he was one of the greatest self-taught illusionists the world had ever seen, presumably using an astonishing combination of sleight of hand and suggestion to achieve his outstandingly theatrical ends.

The results of the Mornington Road experiments appeared to be extraordinary. If they were true, then they contradicted established laws regarding the conservation of energy. In the previous century, Schopenhauer had speculated that, in incredibly rare circumstances, humans could experience supernatural manifestations of the force he referred to as 'will', and the 'sensitive' subjects of Reichenbach had seen odic force in dark rooms. Cromwell Varley's interest—which had inspired William Crookes—was in the nature and detection of fields of electrical force (initially inspired by Faraday and, later, Maxwell), their effective boundaries and their relationship to the mind. In the next phase of his investigative work, Crookes determined to show that psychic force could manifest itself as a physical, transformative phenomenon. For a test subject, he turned once again to a spirit medium.

Since the early 1850s, American mediums had been visiting and touring Britain, inspired by the success of Kate and Margaret Fox. The phenomena generated in the Séance room were interpreted in a range of ways, from telepathy and telekinesis as aspects of spiritual science, to spirit voices and messages attributed to a religious re-awakening. As we have seen, technological breakthroughs such as the Atlantic Cable were often presented as miraculous events inspired by divine agency. Documents such as the *Report on Spiritualism* reveal a range of remarkable conclusions regarding the data recorded, with one correspondent asserting that such activity could not be the work of God, but might be some resurgence of magical, pagan energy.[2] What was certainly true of William Crookes was that—as an investigator from outside the academic elites of the time—he held both a commitment to, and a passion for, addressing questions that perplexed the general public. He undertook such investigations boldly, regardless of the risk to his reputation, and he was also prepared to accept multiple interpretations of his findings. This open approach frequently resulted in clashes with rivals who sought to standardize experimentation,

[2] *Report on Spiritualism*, page 254.

or to assert a broadly materialist world view that restricted imaginative speculation.

It's not clear at what point Crookes began to attend the séances of Florence Cook but, by the time of his investigation into Home, he was certainly participating in meetings in which Cook was the focus of activity. Born in 1856 and raised in Dalston, north-east of central London, Florence Cook was something of a teenage prodigy. What was particularly remarkable about her séances was her apparent ability to manifest spirits as fully formed bodies. In the early years of spiritualism, mediums had apparently manifested hands in the séance room, but the majority of phenomena were aural—knocking (spirit rapping), voices and music. This made sense to scientifically minded spiritualists because it reminded them of the development of Morse code for the purpose of sending messages across the Atlantic via the telegraph cable. The code worked through the opening and closing of electric circuits for different lengths—as dots and dashes—which could be interpreted as letters. To some, the possibility of mental signals 'transmitted' from the brain into the air seemed plausible. However, the new iterations of mediumship, mostly practised by young female mediums, would become increasingly visual and dramatic.

Florence Cook offered an exciting range of new and mysterious interactions with spirit power. She claimed to be a materialization medium, able to realize hands, limbs and, finally, bodies in the séances room. Her spirit form, Katie King, had become a celebrity in her own right, claiming to be the daughter of Henry Owen Morgan, a reformed seventeenth-century pirate who became Governor of Jamaica. But these exotic and hugely popular appearances aroused suspicion and jealousy. A spiritualist named William Volkman took matters into his own hands, seizing Katie during a séance on 9 December 1873 and finding her to be quite solid. As his letter in the spiritualist newspaper *The Medium and Daybreak* one week later stated: 'her ghost-ship could not release her fingers from my hand'.[3] Spirit and sceptic proceeded to engage in a frantic tug of war that culminated in Katie struggling beyond the curtain 'aided into her cabinet by a Justice of

---

[3] *The Medium and Daybreak: A Weekly Journal Dedicated to the History, Phenomena, Philosophy and Teachings of Spiritualism*, 19th December 1873. http://iapsop.com/archive/materials/medium_and_daybreak

the Peace'.[4] Matters concluded when: 'the gaslight was extinguished...and the medium was found in the cabinet five minutes later in some distress'.[5]

The events of 9 December demonstrated that the séance had developed (or perhaps degenerated) into an elaborate form of performance and mediumship suffered a near-fatal crisis of credibility as a consequence. The séance, for mediums such as the British-born American Emma Britten, had been a means of expressing complex and challenging ideas via an intelligent negotiation of voices. Her séances known as 'The Winter Soirees' and given in Harley Street in 1866, displayed a powerful engagement with matters of physics and astronomy as she discoursed on contemporary scientific matters whilst in a trance. By the 1870s, a shift in the method of presentation had occurred. Florence Cook's séances were popular events—valued as both demonstrations of psychic power and as spectacle—with places at sittings highly prized. And there was much to see, even if the eye was straining in the gaslight. Katie King was a ravishing and seductive spectacle, as a *Daily Telegraph* correspondent found: 'a tall female figure draped classically in white, with bare arms and feet...stood statue-like'.[6] The whole scene was exotic: 'we opened the door, and from it suspended a rug or two...in the fashion of a Bedouin Arab's tent'.[7] The séance had become an elaborate and enthralling form of visual performance. But it had also become deeply questionable as the subject of respectable scientific enquiry.

The new generation of mostly female mediums had developed a wide range of skills to engage and deepen their relationship with Victorian audiences. This group included Catherine Wood, Annie Fairlamb and Florence Cook's younger sister, Kate Selena. Their success led to an even more intense scrutiny regarding the validity of spiritualism and séance. For many investigators, including the eminent Cambridge-based mathematician Eleanor Sidgwick, the more performative elements in séance were sometimes problematic. In the words of her niece, Ethel Sidgwick: 'the subject of Spiritualism had been banned, broadly speaking, by the professional scientist. One lost caste by taking it seriously'.[8] In 1875, Eleanor Sidgwick's observations of Catherine Wood led her to a damning conclusion: 'The

[4] Ibid.
[5] Ibid.
[6] *The Daily Telegraph*, 12th August 1873.
[7] Ibid.
[8] Ethel Sidgwick, *Mrs Henry Sidgwick: A memoir by her niece*, (London: Sidgwick and Jackson, 1938) page 90.

indications of deception were palpable and sufficient'.[9] As Emma Britten would conclude in *Modern American Spiritualism*:

> The most severe blows that Spiritualism has sustained have been those aimed by unprincipled and avaricious mediums who, when the manifestations failed to come...practised imposition to supply the deficiency.[10]

Britten attempted to anchor mediumship respectably as both a spiritual *and* scientific practice, suggesting that it was a power that was mysterious, but accessible in all people. The soul 'can pass like thought or electricity with inconceivable rapidity through space'.[11] But Britten, studious, humble and ideas-driven, was fighting against the prevailing trends towards the theatrical, and the credibility of the séance could not survive the shift towards visual manifestation. Spirit voices streaming through the air might be acceptable, young women dressing up as ghosts were not.

The success of the mediums of the 1850s and 1860s owed much to their ability to serve as conduits to a range of supposed spirit voices. Daniel Dunglas Home would enter into a trance state and allow a spirit called Dan to speak through him, often undertaking hazardous tasks such as holding hot coals under the spirit's influence.[12] Home did not always perform in darkness and several witnesses in the Report on Spiritualism stated that they had witnessed feats of mediumship in full light.[13] But, as Antonio Melechi has suggested, Home's advantage was 'the respectable milieu in which he exclusively operated. Never scrutinizing his motives or methods, polite society gave Home the latitude to dictate the conditions in which he worked'.[14]

The materialization séance, when successful, combined a number of key elements, including use of the spirit cabinet and the presentation of an extraordinary and exotic figure. Florence Cook's control spirit, Katie

---

[9] Ibid., page 97.
[10] Emma Hardinge, *Modern American Spiritualism*, (New York: The Author, 1870) page 337.
[11] Emma Hardinge, *Address Delivered at the Winter Soirees*, (London: Thomas Scott, 1866) page 17.
[12] *Report on Spiritualism*, page 207.
[13] Ibid., page 215.
[14] Antonio Melechi, *Servants of the Supernatural: The Night Side of the Victorian Mind*, (London: William Heinemann, 2008) page 206. This, it must be said, stands in contrast to the laboratory tests of 1870 conducted by Crookes.

King, was just such a figure. It was her appearance out of the darkness that was the most important aspect of the séance, and the main attraction to Florence Cook's public. Usually, the medium was secured in the cabinet and a hymn was sung, the gas light was lowered, and the spirit achieved physical realization in the gloom. Key to the authenticity of the manifestation was the fact that the body of the medium was secured in the cabinet whilst the spirit form appeared outside of it. If not a miracle of mediumship, then the materialization séance was a highly developed escapology act.

Although popular in the 1870s, the first materialization séances had occurred as early as 1861 in New York and were conducted by the 25-year-old Kate Fox. Fox had apparently materialized a spirit at the request of a grieving banker named Charles Livermore, who had lost his wife, Estelle, in 1860. This event created the template for the materialization séance that others, including Florence Cook, would seek to emulate. According to author Barbara Weisberg:

> When the manifestations began, they more than fulfilled his (Livermore's) expectations. On October 20th, 1861, Estelle stood in front of him, enveloped in her gossamer robes, her arm bare except for its transparent drapery.[15]

The sittings became increasingly emotional; 'before returning to the spirit world, she (Estelle) placed her finger enveloped in gossamer several times in his mouth, an erotic and intimate gesture'.[16] Weisberg writes further of the theatricality of this spectacle, which held Livermore and his doctor, John Gray, in total thrall: 'Livermore often used language suggestive of a magic lantern show to evoke what he saw: for example, he described how the spirit light on one particular evening rose in a cloud'.[17] Livermore went on to describe how the face and figure of his wife were projected 'with stereoscopic effect...We were told to notice her dress, which seemed tight-fitting'.[18] It was an extraordinary spectacle and demonstrated the dramatic power of a successful materialization séance, with medium and figure seemingly both present in the room at the same time. The documentation of such a double appearance—medium and spirit together—by photography or witness statement and verified using the latest electrical

---

[15] Barbara Weisberg, *Talking to the Dead* (New York: Harper Collins, 2004).
[16] Ibid.
[17] Ibid.
[18] Ibid., page 203.

technology, would be the principal research aim of William Crookes in his laboratory tests of Florence Cook.

After the disastrous séance of 9 December, Florence Cook submitted herself for laboratory tests to be undertaken by William Crookes. Like many private mediums, Cook was financially supported by her local spiritualist association, and a wealthy patron called Charles Blackburn. Her professional survival depended upon such backing and it's likely the connection with Crookes was an attempt to restore credibility with her patron. In February 1874, William Crookes announced his intention to test Florence Cook in order to 'give the weight of my testimony in favour of her whom I believe to be unjustly accused'.[19] Crookes had been invited to a séance shortly after the Volkman incident at the home of the spiritualist John Chave Luxmoore. While at this event, Crookes claimed to have seen the spirit form Katie King in the séance room whilst hearing 'a sobbing, moaning sound come from behind the curtain'.[20] The curtain was an improvised spirit cabinet set up in Luxmoore's home that allowed for the medium to be entranced and restrained in the dark so the spirit form could be released or projected to move among the sitters. Key to the authenticity of the materialization séance was the existence of two distinct forms: spirit body and 'magnetized' or entranced medium. The author Trevor H. Hall—a critic of Florence Cook's mediumship[21]—has pointed out that Florence spent time with the mediums Herne and Williams, both of whom were caught in cases of outright fraud. As Hall states, their 'training' of young mediums included the production of vocal phenomena. Yet Crookes was clearly intrigued by this encounter with Katie King and determined to proceed with his investigation. He announced: 'I am promised that every desirable test shall be given to me'.[22]

Two things immediately stand out in a consideration of Crookes' approach to the investigation of Florence Cook. The first is the lack of named and credible witnesses. Although he did not name William Huggins and Edward Cox in his first report on the Home séance in the *Quarterly Journal of Science*, he did give their qualifications and their identities were revealed via the published correspondence which supported Crookes'

---

[19] William Crookes, page 102.
[20] Ibid., page 103.
[21] Hall, T.H. *The Medium and the Scientist: The Story of Florence Cook and William Crookes* (New York: Prometheus Books, 1984).
[22] William Crookes, p. 106.

astonishing description of what Home had apparently achieved. The second is the lack of clarity regarding where the investigations were taking place. With the testing of Home, Crookes was clear from the start that the activity was *only* taking place in his Mornington Road laboratory. But the experiments with Florence Cook—reported first in the spiritualist press—lack specific location and appear to take place in the laboratory *and* in the homes of spiritualists. As scientific investigations, they were compromised and Crookes would have known it. One of the most remarkable aspects of the Home investigation was the phenomena witnessed by the lab assistant Charles Gimingham—a veteran of many scientific demonstrations. Yet it's unclear if he was present at any of the Florence Cook experiments.

For the Cook séances, Cromwell Varley proposed making use of a new method of electrical control that he had devised. Varley proposed to place the medium in a circuit with a two-cell battery, two sets of resistance coils and a 'galvanometer'.[23] The galvanometer would be visible outside of the cabinet (or curtained area depending on the venue of the séance). A mild electric current would pass through the medium and, if she broke the circuit by movement, the galvanometer would reveal it. At some point in March 1874, in a scene that was strangely theatrical and somewhat reminiscent of Gray's use of the Hauksbee Machine in the previous century, Florence Cook took her place on a seat in a darkened cabinet. According to a report in *The Spiritualist* newspaper, pieces of blotting paper 'moistened with a solution of nitrate of ammonia'[24] were attached to her arms and coins were placed upon the blotting paper, with paper and coins held in place by 'elastic bands'.[25] Wires were attached to the coins and then to the resistance coils and galvanometer. Varley observed the galvanometer for evidence of movement and the breaking of the circuit that might reveal Cook to be a fraud. The unnamed sitters sang a hymn and settled down to wait for the manifestation of the spirit Katie King. She appeared 28 minutes into the sitting, with Varley reporting significant movement 'a fall of thirty-six divisions' but 'no break of circuit'.[26] Katie continued to interact with the sitters whilst Florence, supposedly, remained entranced in the chamber.

---

[23] For measuring the strength of electric current.
[24] *The Spiritualist,* 20th March, 1874, page 133, http://iapsop.com/archive/materials/spiritualist/spiritualist
[25] Ibid.
[26] Ibid.

The *Spiritualist* newspaper was confident in its conclusion that the experiment proved the genuineness both of Florence Cook and the practice of materialization mediumship. Varley the great electrician and Crookes the great chemist had apparently confirmed the validity of the spiritualist project. Varley's test was employed to 'verify' the claims of a number of mediums in the subsequent years, with the *Medium and Daybreak* also publishing a blueprint for the experiment in an article entitled 'A Scientific Séance – The Electrical Test for Mediumship' in 1875.

So what had happened? Why were the controls that had been meticulously put in place to test Daniel Dunglas Home ignored, overlooked or, at best, not documented in the investigation of Florence Cook? Why was such an elaborate electrical test set up when Edward Cox had already suggested several methods to ensure the spirit form and the medium were separate entities: leave the cabinet door open or place a sitter within it, or strap the medium to the chair in a way that would not allow the unpicking of knots? (The séance of the Davenport Brothers—rivals to the Fox Sisters from nearby Buffalo, NY—had their cabinet act exposed and eventually sold their escapology secrets to Houdini.) Several theories have been advanced, from Trevor H. Hall's proposal that Crookes and Cook were involved in a sexual relationship to that proposed by the anthropologist Edward Clodd that Crookes' eyesight was deteriorating so rapidly that he was easily duped, and too proud to admit it when the fraud became clear.[27] Crookes' attraction to mediumship was a complex matter, and his investigations demonstrated a blurring of personal and professional interests. In purely practical terms, the electrical test protected the 'sensitivity' of the medium, as well as her privacy. These gendered concerns around the idea of female 'frailty' would have undoubtedly mattered to Crookes and, whatever the truth of any alleged romantic relationship with Cook, he would have likely wanted to preserve a sense of propriety and to maintain the appearance of chivalry. As we have seen in the Home investigation, it's likely that the loss of his brother, and the failure of the subsequent lawsuit, might have provided an urgent, emotional motivation for his enquiries into spiritualism. This may have led him to drop his guard during his interactions with Florence Cook.

Daniel Dunglas Home retired from giving séance after the extraordinary events in the Mornington Road laboratory in 1871. Florence Cook

---

[27] Edward Clodd, *The Question: A Brief History and Examination of Modern Spiritualism*, (London: Grant Richards, 1917) page 100.

made the same decision in 1874. Her final, private, séances took place in Hackney and were reported by *The Spiritualist* newspaper. They increasingly resembled romantic dramas. Of the 13 May sitting, a reporter wrote:

> the whole company were invited to crowd around the door whilst the curtain was withdrawn and the gas turned up to the full, in order that we might see the medium in her blue dress and scarlet shawl, lying in a trance upon the floor.[28]

Crookes invited several notable spiritualists (perhaps he had given up on inviting scientists) to serve as witnesses to Katie King's final appearances. One of these was Enmore Jones. Jones reported his concerns in *The Medium and Daybreak* in the edition of 22 May 1874. He stated the following: 'The leader (Crookes) stood in front of the awning and made himself very active every time that Katie appeared'.[29] Crookes reminded Jones of a 'fussy mesmerizer' and was 'half-showman and half-playactor'.[30] On 21 May 1874, Florence produced Katie King for the final time, and the spirit's departure to the next world did not disappoint her devotees. In the 29 May edition of *The Spiritualist*, Florence's supporter Florence Marryat described the séance in the following terms:

> Katie was very busy that evening. To each of her friends assembled to say goodbye, she gave a bouquet of flowers...a piece of her dress and veil, and a lock of her hair...She wrote: 'From Annie Owen de Morgan (alias Katie King) Pensez a moi'.[31]

William Crookes wrote of his final encounter with Katie:

> After closing the curtain she conversed with me for some time, and then walked across the room to where Miss Cook was lying senseless on the floor. Stooping over her, Katie touched her and said: 'Wake up, Florrie, wake up! I must leave you now'. Miss Cook then woke and tearfully entreated Katie to stay a little time longer.[32]

---

[28] *The Spiritualist*, 29th May 1874, page 258.
[29] *The Medium and Daybreak*, 22nd May 1874, page 621. http://iapsop.com/archive/materials/medium_and_daybreak
[30] Ibid.
[31] *The Spiritualist*, page 134.
[32] William Crookes, page 111.

This report ensured that Crookes, to quote Edith Sidgwick, 'lost caste' and credibility amongst some members of the scientific community. He would, in time, recover his reputation and become a part of the intellectual establishment of the late-Victorian period, but his involvement with this aspect of spiritualism probably compromised the progress of his scientific career in some areas. For his investigations in the field of spectroscopy and on the nature of cathode rays, William Crookes ought to be better-known and celebrated. His reasons for departing from the laboratory method employed with Home in his investigation of Florence Cook cannot be entirely understood by us a century and a half later. As I have mentioned, Crookes was another scientist whose early career was shaped outside the Victorian elites. He felt an obligation to the public to investigate issues of concern and was influenced in that mission by Michael Faraday. Crookes viewed his role in similar terms to that of his mentor, as a teacher and interpreter of scientific phenomena to the public. The work of secular, investigative groups such as the London Dialectical Society probably placed a kind of critical mass upon Crookes that compelled him to explore some distinctly exotic occurrences. Whatever his later motivations, it seems likely that from the death of his brother in 1867, Crookes openly and genuinely sought to explain what many of his fellow citizens were reporting as facts: spirit rapping, voices and other apparent manifestations of mysterious forces. Regarding the differences in his approach to Home and Florence Cook, these were largely conditioned by gender-specific protocols. As a gentleman, it was acceptable for Crookes to watch Home dress at his apartment before scientist and medium departed for the Mornington Road investigation. Obviously, this could not have happened with Florence Cook. Exposure of female mediums—usually by their rivals—generally showed them to be concealing their ghostly costumes within their crinolines, or even within their underwear.

Victorian concepts of respectability and decency probably meant that Crookes placed a great deal of trust in Florence Cook, seeing her as a young and vulnerable innocent in need of his protection. It's impossible to say whether or not this chivalric model became something else over time, but it seems likely that Florence Cook was able to use gendered concepts of frailty and sensitivity to prevent the kind of thorough investigation that her claims required. But what did Crookes believe he was witnessing in her séances? He seems to have believed that a medium's spirit form could be a form of 'radiance', a manifestation of 'higher' matter in a new form that had been predicted by Faraday in the decades before. This matter might need to 'develop' in specific conditions and require darkness. Such a proposal would explain why Crookes permitted Cook to make use of the darkened cabinet. If psychic force was related to the odic force of Reichenbach, then it would demonstrate luminosity around the body of the medium. In this respect, this investigation makes a kind of logical sense as part of Crookes' research programme. Psychic force might manifest itself in various fields and across frequencies, generating a range of different phenomena. Alternatively, Crookes may initially have considered the image of Katie King to be a kind of telepathic experience, a sharing of the medium's mind and sensory data with the sitters, again based upon some concept of an active field. This would make sense when considered alongside some of the observations of the work of Home made by members of the London Dialectical Society. But nothing can rationally explain Crookes' claim to have observed a ghost embrace her medium and bid her farewell, or the sharing of locks of her hair and personal notes. These later observations, made entirely outside of the laboratory criteria established for Home, indicate probable complicity with the medium, or an outstanding level of naivety on the part of the scientist. His motivations are, of course, as mysterious to us as the phenomena spiritualists claimed to have experienced in the séance room. We might conclude that, at some point, the grieving William Crookes fell under the spell of a brilliant, improvising magician. This led him to abandon the programme of research he'd entered upon, and to relax the investigative framework he'd previously established with Home. Some scientific spiritualists were already dubious about materialization mediums, and aware of the fact that the electrical test could be manipulated or misread via the careful substitution of elements in the circuit. Whatever the true nature of the relationship between Crookes and Cook, we shouldn't entirely dismiss the rigour of the scientist. His work prepared the ground for significant advances in chemistry and physics. From the 1850s, his studies involved close examination and notation of complex data, including weather systems during his early research work at Chester Diocesan College. His research not only identified the element thallium but also led to the calculation of its atomic weight.

It's likely that he believed that certain types of force could be generated by the mind, but only in unique and very specific circumstances. Like Elliotson, Crookes didn't claim that his experiments were repeatable because they dealt with human subjects who were susceptible to a range of influences. Both animal magnetism and psychic force were potentially present in all people, but only realized in a few. It's likely that the challenge of investigating mediums attracted Crookes because some of the concepts involved were, distantly, connected to ideas put forward by his hero, Michael Faraday. Crookes' early experiments were, possibly, testing the boundaries of the Kantian concepts of noumenon and phenomenon. If human perception is a limited band of perceptual wavelengths, then is that a constant, or could it vary from person to person, and how drastically? Although we may view the investigation of both mediums as strange or even comical, Crookes' investigations—particularly of Home—were refreshingly free of prejudice, and as radically experimental in their way as the work that yielded the discovery of thallium. As we shall see, there's a very clear difference to be observed in the practical, person-centred experimentation of Crookes and the more theoretical and distant approach favoured by the rather more patrician academics who formed the Society for Psychical Research in 1882.

In his biography of Crookes, published in 1923, E E Fournier D'Albe offered two responses to the mystery of the scientist's interest in séances. The first is titled 'The Spiritualist Version'. In this version, the author asserts that the experiments with Home were 'classical and absolutely free from flaw',[33] and that those with Florence Cook 'convinced him that there was supra-mundane intelligence behind the phenomena, and so he became a convinced spiritualist'.[34] The second version, titled 'The Rationalist Version', offers an image of Crookes as a man more laudable and chivalrous than those he was investigating. According to this version, he 'fell an easy victim to the impostors who were then ministering to what had become a society craze in England…he was deceived by both Home and Florence Cook'.[35] He came to realize that 'science and Spiritualism were incompatible and incommensurate, science being of the mind and Spiritualism being of the heart'.[36]

D'Albe may well have hit upon the truth when he referred to Home's intense rivalry with the scientist W B Carpenter. At the time of the Home investigation, Carpenter was Registrar of the University of London. His

---

[33] E. E. Fournier D'Albe, *The Life of Sir William Crookes*, (London: T. Fisher Unwin Ltd., 1923) page 174.
[34] Ibid., page 175.
[35] Ibid., page 176.
[36] Ibid., page 178.

letter to the London Dialectical Society in December 1869 contained a damning critique of the phenomena associated with spiritualism:

> My enquiries have led me to the conclusion…that the source of these phenomena does not lie in any communication ab extra, but that they depend upon the subjective condition of the individual which operates according to certain recognized physiological laws. These I expounded in an article on mesmerism, electro biology etc. which I wrote for the Quarterly Review in October 1853; and I have not seen any subsequent reason for modifying the opinions therein expressed.[37]

Crookes named Carpenter as the anonymous author of another article in the *Quarterly Review* of October 1871. The article attacked Crookes, Huggins and Cox as unreliable witnesses in terms that are remarkably personal and indiscreet by Victorian standards. One by one, their credentials for examining such phenomena are demolished. Crookes, never one to shy from an academic contest, was quick to marshal his defence. He recalled an earlier encounter with Carpenter in Edinburgh some years before:

> I had an opportunity of observing the curiously dogmatic tone of his mind and of estimating his incapacity to deal with any subject conflicting with his prejudices and prepossessions.[38]

D'Albe declines to decide in favour of the two versions, preferring to leave the matter in the hands of future scholars. But he does state that 'Crookes never recanted, never wavered, never withdrew'.[39] Perhaps, having ventured the theory of psychic force prematurely, Crookes felt bound to defend it against an academic rival who lacked his discretion and courteous manner. Carpenter's hypothesis, that all psychic phenomena were based on psychological manipulation or unconscious mental action, probably troubled Crookes' optimistic vision of humankind. Such a world view lacked nobility and opposed the vision of a progressive humanity that might evolve spiritually, or psychically, as well as physically. By locking himself into a position as the defender of psychic force, Crookes travelled deeper and deeper into the world of duplicitous mediums, suspending his credulity and risking his credibility. From the summer of 1874, he limited

---

[37] *Report on Spiritualism*, page 266.
[38] E. E. Fournier D'Albe, page 213.
[39] Ibid.

his investigations in this field. Perhaps his partial withdrawal and Katie King's 'departure' from the mortal world were not entirely coincidental. As Richard Noakes has observed, the contest between Crookes and Carpenter was a complex one. It was about different ideas of power and the authority of the séance event. Ultimately, as Noakes writes: 'What was at stake were rival notions of the scientific, the natural, the lawful'.[40] In many respects, their struggle is reminiscent of the earlier rivalry between Elliotson and Wakley, with both parties *absolutely* committed to their respective interpretations of the data and allowing no room for compromise. Ultimately, the life of psychic force was brief, yet its impact on the imaginations of some of Crooke's associates would prove to be profound.

---

[40] Richard Noakes, 'Spiritualism, science and the supernatural in mid-Victorian Britain' *The Victorian Supernatural*, eds. Bown, Burdett and Thurschwell (Cambridge: CUP, 2004) page 39.

CHAPTER 9

An Eternal Mind

Abstract In this chapter, I examine developing ideas of the mind in the late nineteenth century, focusing on the investigations of physicist Oliver Lodge. Lodge hypothesized that the mind might exist beyond the constraints of the physical brain. This resulted in a range of experiments in the field of telepathy. At the same time, the Society for Psychical Research (SPR) was formed by scholars in Cambridge, dedicated to examining the credibility of such phenomena. I consider the tensions between spiritualists and the SPR, and the increasing divide between the investigation of occult forces and 'legitimate' scientific practice.

Keywords Telepathy • Psychic • Ether • Mind • Sensitive • Society for Psychical Research • Oliver Lodge • W B Carpenter • Eleanor Sidgwick

> I wish to make the hypothesis that it is the Ether which is really animated, and that this animated ether interacts with matter; I suggest that the true vehicle of life and mind is Ether, and not matter at all (Oliver Lodge, *Ether and Reality*, 1925).[1]

---

[1] Oliver Lodge, *Ether and Reality: A Series of Discourses on the Many Functions of the Ether of Space*, (Cambridge: CUP Library Collection, 2012) page 166.

In June 1894, the British scientist Oliver Lodge demonstrated a small cable transmitter at London's Royal Society. Using Morse code—previously only used for messages via the Atlantic Cable—Lodge sent short pulses of electricity through the air that were received by an operator in another room. Lodge had successfully demonstrated radio telegraphy. Independently of Lodge, in Pontecchio, Italy, Guglielmo Marconi made a bell ring across a room via a transmitter. Marconi's practical demonstrations soon expanded further as he experimented with antennae to send messages over greater distances. In the summer of 1895, Marconi sent messages from his family's garden to a receiving station two miles away. The waves he sent didn't just pass through air, they passed through forests and even through brick walls. Quite independently of each other, Lodge and Marconi had proved that electro-magnetic waves could operate independently of expensive infrastructure. To many, it must have seemed even more miraculous than the messages celebrated in the religious sermons of the 1850s that eulogized the first iteration of the Atlantic Telegraph.

Both experiments were built on the discovery of the German physicist Heinrich Hertz, who had proved the existence of Maxwell's proposed electro-magnetic waves in 1886 and put them to an extraordinary practical use via the 'tapper' devices employed at the signalling stations for the Atlantic Cable. Lodge didn't patent his device, but Marconi did and achieved great recognition and financial success for this remarkable technological breakthrough. Lodge, like William Crookes, was a scientist whose openness to diverse areas of scientific enquiry and defiance of concepts of intellectual respectability may have led to his partial-exclusion from the list of most-notable Victorians. Of all the scientists of the period, Lodge had arguably the greatest vision of all, for he sought not only to demonstrate the mysterious forces at work in the world, but also to fit them into an all-consuming scientific concept. He called this 'the ether'.

The term had, of course, been in use for centuries. Its classical iteration was the mythic substance breathed by the gods of Olympus in the Greek pantheon. At the time of Newton's theory of gravity, ether was a hypothetical substance that filled the area of space beyond the earth's 'terrestrial sphere'. It provided a means by which light could travel from the sun and was imagined as a conduit—possibly as another kind of fine, invisible fluid—the medium light was travelling through and interacting with. Lodge proposed it as the binding or unifying force within the universe, a substance from which particles received mass and became realized as phenomena. Lodge—like Orsted and Faraday before him—sought to reconcile the troubling Kantian impasse of noumenon and phenomenon via the concept of gradations of fineness in the etheric material. All was a continuum and endlessly connected, if only the scientist could provide evidence for this mysterious matter.

Like so many in this story, Lodge took risks with his well-being and reputation in his quest to discover the structures that underpinned reality. His father worked for the railways before joining the growing pottery industry in Staffordshire. He became a clay merchant, travelling to the south-west of England to source materials. He withdrew Oliver from school aged 14 so he could join him in the family business. Lodge junior hid any resentment in his biography *Past Years*, published in 1931. He recounted making regular journeys to London and staying with his Aunt Anne in Fitzrovia. She saw his intellectual potential and encouraged him to attend the free lectures given at the Royal Institution, still presided over by Michael Faraday. In *Past Years*, Lodge described the effects of these scientific lectures in near-spiritual terms:

> I have walked back through the streets of London or across Fitzroy Square with a sense of unreality in everything around, an opening up of deep things in the universe, which put all ordinary objects of sense into the shade, so that the square and its railings, the houses, the carts and the people, seemed like shadowy unrealities, phantasmal appearances, partly screening, but partly permeated by, the mental and spiritual reality behind.[2]

The sense of being overcome by the idea of a vast and invisible world was familiar to the generation of scientists working in the wake of Kant's concept of the noumenon. Like others in this tradition, Lodge's mind was a blend of the practical and the romantic. Thanks to the mentorship of the physicist George Carey Foster, a professor who became a friend and supporter of his career, Lodge was able to work as a demonstrator at University College, eventually rising to the position of assistant professor. He remained in London until 1881, when he accepted a position as Professor of Physics at University College Liverpool.

Over 40 years later, Lodge attempted to bring the threads of his vast range of interests together into an accessible work known as *Ether and Reality*. His description of the ether reflected his ambition and that of so

---

[2] Oliver Lodge, *Past Years: An Autobiography*, (London: Hodder and Stoughton, 1931) page 78.

many of his contemporaries, whilst also honouring the esoteric interests of Newton: 'It is the primary instrument of Mind, the vehicle of the Soul, the habitation of Spirit. Truly it may be called the living garment of God'.[3] This language might strike us as unscientific, yet it is indicative of the exotic blend of faith, philosophy and scientific enquiry that informed much research during this period. It's worth noting that Albert Einstein first encountered the works of Kant in 1892, aged just 13. The quest to solve the problems of philosophy motivated and inspired many young working people to involve themselves as scholars and amateur experimenters, or to become self-improving readers via the burgeoning scientific press. The range of interests addressed by this press often troubled scientists such as W B Carpenter who were attempting to remove the more arcane and wilder elements from the discourse. But this quest for respectability shouldn't eclipse the scale of the thought that men and women of this period were capable of. What concerned them—even obsessed them—were the biggest questions of life and death.

Lodge's theories about the ether were inspired by the discoveries of James Clerk Maxwell. Like Alfred Russel Wallace, he applied them to human consciousness, asking if the mind could be interpreted as an electrical 'field' distinct from the physical restrictions of the brain, perhaps in some way analogous to the process at work in wireless transmission. This had initially been termed 'thought transference', then 'thought reading' before the investigator Frederic Myers applied the neologism 'telepathy'. John Elliotson believed he'd detected evidence of a similar phenomenon in his experiments at University College Hospital with the Okey sisters, who appeared to have sent thoughts to each other whilst in a mesmerized or 'magnetic' state. Lodge had initially been sceptical of these accounts, but when a Justice of the Peace contacted him to report the actions of two young 'psychics' in Liverpool, Lodge felt compelled to investigate.

Fortunately, there was a new, respectable organization to give weight to such enquiries: the Society for Psychical Research. Formed in February 1882, the Society's founding members were, predominantly but not exclusively, a group of intellectuals from Cambridge. Since the London Dialectical Society's investigation of spiritualism, groups of interested citizens and scholars had continued to assess the credibility of the séance and its attendant phenomena. Among them was the formidable mathematician Eleanor Balfour, sister of the politician (and eventual prime minister)

[3] Oliver Lodge, *Ether and Reality*, page 179.

Arthur Balfour, who married the philosopher Henry Sidgwick in 1876. As Eleanor Sidgwick, she investigated a number of materialization mediums and, as we've already seen, she was largely unimpressed by their work. In 1886, as an increasingly influential investigator for the SPR, she exposed the medium William Eglington as a fraud by demonstrating that his use of 'spirit plates' was simply a magic act that could be replicated. It's hard to imagine that Sidgwick would have allowed Florence Cook the degree of freedom permitted to her by William Crookes and Cromwell Varley in the test séances of 1874, and her aggressive investigative approach led to the alienation of many spiritualists. In the second volume of *The History of Spiritualism*, published in 1926 and dedicated to Oliver Lodge, Arthur Conan Doyle didn't hold back in his criticism of the SPR and its 'strangely mingled record of usefulness and obstruction', accusing its leadership of 'a certain supercilious air towards Spiritualism'.[4] He singled out Eleanor Sidgwick for particularly harsh criticism, claiming that she was 'one of the worst offenders' in making accusations of fraud towards mediums. Conan Doyle reported how: 'One lady medium, the daughter of a well-known professor, described to the author how impossible, and indeed how unconsciously insulting, was the attitude of Mrs. Sidgwick on such an occasion'.[5]

In contrast to many of the leading scientists of the time—including former practical demonstrators Crookes, Russel Wallace and Lodge—the founders of the SPR were, broadly speaking, members of the Victorian social elites who had the time and resources for sustained investigation. The Sidgwicks were joined by the scholars Frederic Myers, William Barrett, Edmund Gurney and Lord Rayleigh. Acutely aware of losing credibility, the SPR leadership was sceptical about the more theatrical manifestations of séance and spiritualist practice and focused instead on psychic phenomena of a more demonstrable and testable kind, what Conan Doyle termed 'mental' phenomena.[6] William Barrett had already presented a paper on 'Thought Reading' in 1876 and called for the formation of an academic committee to investigate the matter further. The SPR emerged from a collective desire to go beyond the limitations of conventional enquiry and to examine new phenomena related to ideas of force and matter. But it

---

[4] Arthur Conan Doyle, *The History of Spiritualism*, Vol. 2, (London: Cassell and Company, 1926) page 55.
[5] Ibid., page 57.
[6] Ibid., page 58. It should be noted that SPR members investigated the highly theatrical phenomena demonstrated by Eusapia Palladino in the 1890s, in Cambridge and in France.

guarded its credibility very carefully. As its founding statement made clear, it was set up to investigate 'mesmeric, psychical and spiritualist phenomena in a purely scientific spirit'.[7] As a respectable scientific and philosophical body with a clear mission, the SPR was able to acquire the skills and knowledge of a range of scientists. Oliver Lodge was outside of the Cambridge-based elites, but his concept of the ether was an attractive idea that potentially bound a diverse set of phenomena together.

The SPR's first investigations of telepathy were published in the organization's Journal and collected in the two-volume work *Phantasms of the Living* in 1886. The authors are given as Edmund Gurney, Frederic Myers and Frank Podmore, although other investigators, including Oliver Lodge, are referred to throughout. The authors report the link between mesmerism and telepathy that Elliotson had related in his investigations at UCH, but they are also clearly keen to reboot the investigation without the theatrical taint of Mesmer and other 'extravagances'.[8] Barrett is positioned as the first truly credible discoverer of this new ability when he 'brought under the notice of the British Association at Glasgow a cautious statement of some remarkable facts which he had encountered, and a suggestion of the expediency of ascertaining how far recognised physiological laws would account for them'.[9] From the outset, the investigative team were quick to lay down credible and respectable parameters. Indeed, the principal critic of such endeavours—William Carpenter—was directly appealed to, but Carpenter was unconvinced by the evidence presented so far and remained sceptical. Always sensitive to materialist criticisms, the SPR investigators were careful to detail at length the checks and conditions for each of their experiments. Carpenter had explained most 'paranormal' ideas as variations of ideomotor action—unconscious movement and signalling by the body during the experiment. Barrett had practised a party trick known as 'the willing game' in which a person leaves a room, and the remaining guests hide an item. The returning guest, termed the 'percipient', has another guest take them by the hand, or place a hand on their shoulder. The 'willer' or guide unconsciously gives bodily signals and reassurances as the percipient draws close to the item, confirming Carpenter's argument:

[7] https://www.spr.ac.uk/about
[8] Edmund Gurney, Frederic W. H. Myers and Frank Podmore, *Phantasms of the Living* Vol. 1 (London: Trubner and Co.) page 13.
[9] Ibid., page 14.

till this game was played, probably no-one realised that muscular hints, so slight as to be quite unconsciously given, could be equally unconsciously taken, and that thus a definite course of action might be produced without the faintest idea of guidance on either side.[10]

And there were other ways that information could be relayed:

In some cases it appeared that even contact could be dispensed with and the guidance was presumably of an auditory kind – the 'subject' extracting from the mere footsteps of the 'willer', who was following him about, hints of satisfaction or dissatisfaction at the course he was taking.[11]

To pre-empt further criticism, the SPR team outlined their awareness of the possibilities of conspiracy between the test subject and the examiner, giving Morse Code as a possible method:

The long and short signs of the Morse Code admit of many varieties of application, and though the channels of sight and touch may be cut off, it is difficult entirely to cut off that of hearing. Shufflings of the feet, coughs, irregularities of breathing, all offer available material.[12]

Gurney, Podmore and Myers wanted to make it clear that they were above being tricked by the kind of 'thought reading' magic acts that were popular in Victorian music halls. Their tests would be of the strictest kind, with no opportunity for fraud, conscious or unconscious. Even Carpenter, they insisted, would be satisfied by the checks put in place by SPR investigators.

Barrett began investigating the Creery family of Buxton, Derbyshire, in 1881. There were few concerns regarding the credibility and respectability of the test subjects: Thomas Creery was a reverend and none of his five daughters, aged between 10 and 17, needed to be placed into a mesmeric state or sleep. These were middle-class girls, raised in faith, and the prejudices often applied to female mediums were not considered relevant in this case. Barrett described a wholesome and strict test held at the family home at Easter 1881:

---

[10] Ibid., page 15.
[11] Ibid.
[12] Ibid., page 18.

> One of the children was sent into an adjoining room, the door of which was closed. On returning to the sitting room and closing its door also, I thought of some object in the house, fixed upon at random, writing the name down, I showed it to the family present, the strictest silence being preserved throughout. We then all silently thought of the name of the thing selected. In a few seconds, the door of the adjoining room was heard to open, and after a very short interval the child would enter the sitting room, generally with the object selected. No-one was allowed to leave the sitting room after the object had been fixed upon, no communication with the child was conceivable, as her place was often changed.[13]

Anxious about the use of children for the 1882 experiments, Gurney went to some lengths to ensure that only the members of the investigative team were able to act as 'transmitters'. The visiting members of the SPR had not met the Creery family and included Gurney, Myers and Frank Podmore. The experiments were moved away from the family home to Cambridge and then to Dublin. The most extraordinary results were achieved using a standard deck of 52 playing cards with strict controls in place: 'The four observers were perfectly satisfied that the children had no means at any moment of seeing, either directly or by reflection, the selected card'.[14] Out of 14 tests using cards, nine 'were successful at first guess, and only three trials can be said to have been complete failures'.[15] The Creery girls were able to reproduce their extraordinary strike rate over successive sittings with other examiners. To dispel an unthinkable suggestion—that the Reverend Creery might be involved in collusion or the author of 'an ingenious family trick'—he was removed entirely from the process on a number of occasions with no change to his daughters' astonishing record of success.[16] Even when the card wasn't guessed first time, the correct suit was named 14 times in succession and 'knave was very frequently guessed as king, and vice versa, the suit being given correctly'.[17] For the investigators, the issue was to explain the nature of the reception of data—was the image 'seen' or was the name 'heard' by the percipient or receiver? Gurney concluded that 'both modes of transference were possible' and that they 'prevailed in turn' depending on the subject's adaptation.

[13] Ibid., page 22.
[14] Ibid., page 23.
[15] Ibid., page 24.
[16] Ibid.
[17] Ibid., page 27.

Lodge was undoubtedly aware of the Creery investigation and had already taken a keen interest in the activities of the stage psychic Irving Bishop in the year before the founding of the SPR. Having moved to Liverpool in 1881 from University College London, Lodge became part of the investigation team when Malcolm Guthrie reported that two young women in his employ at the firm George Henry Lee & Co. had demonstrated remarkable evidence of 'thought reading'. Guthrie contacted James Birchall of the Liverpool Literary and Philosophical Society and together they began to make their own experiments. As Guthrie subsequently wrote: 'The experiments have all been devised by myself and Mr Birchall, without any previous intimation of their nature and could not possibly have been foreseen'.[18] The two young employees, Miss Relph and Miss Edwards, were the subjects of the test. Guthrie and Birchall, as the 'transmitters', drew a sequence of images before 'sending them' as thoughts for the women to reproduce. Guthrie described the method in detail:

> During the process of transference the agent (transmitter) looked steadily and in perfect silence at the original drawing which was placed upon an intervening wooden stand, the percipient (receiver) sitting opposite to him and behind the stand, blindfolded and quite still...the reproductions were made in perfect silence.[19]

The drawings were printed in a section of volume one of *Phantasms of the Living* and showed an extraordinary level of agreement between the two parties. Images including an anchor, a wine glass, a fish and a chair were clearly produced by one or both women under separate examinations. Lodge expanded the range of tests and combined different images in pairs. He also frequently altered the conditions of the tests changing them 'at will'.[20] He stated that the experiments gained in credibility to the point that they were 'quite comparable to that induced by the repetition of ordinary physical experiments'.[21] He asked one of the women what she experienced during the moment of supposed reception:

---

[18] Ibid., page 36.
[19] Ibid., page 37.
[20] Ibid., page 49.
[21] Ibid.

> She said she felt an influence or thrill. They both say that several objects appear to them sometimes, but one among them persistently recurs and they have a feeling when they fix upon one that it is the right one.[22]

Lodge's experiments with the women demonstrated a shift between visual and auditory reception—the form of a thing or the name of a thing—and confirmed Gurney's observation with the Creery sisters that the phenomena diminished over time. Lodge offered a variation on Reichenbach's theory of 'sensitivity' that stressed the 'delicate psychological conditions'[23] of the test subject and drew attention to the significance of the observer in the experiment:

> The man who first hears of thought transference very naturally imagines that, if it is a reality, it ought to be demonstrated to him at a moment's notice. He forgets that the experiment, being essentially a mental one, his own presence – so far as he has a mind – may be a factor in it.[24]

Lodge's tentative hypothesis—that the mind might be analogous to an electrical field—was presented in the *Proceedings of the SPR* in 1884: 'as the energy of an electric charge, though apparently on the conductor, is not in the conductor, but in all the space around it'.[25] Inevitably, this line of thinking led Oliver Lodge back to his ideas about the ether and consolidated his belief that it might act as a kind of host or medium for a range of phenomena. Ultimately—and perhaps fancifully—the ether might challenge the dualism of Kantian philosophy by expressing the noumenon as a range of vibrating forces that could be made knowable in time.

In the *Critique of Pure Reason* of 1781, Kant had asked:

> Now what are space and time? Are they actual entities? Are they only determinations or relations of things, yet ones that would pertain to them even if they were not intuited, or are they relations that only attach to the form of intuition alone, and thus to the subjective constitution of our mind, without which these predicates could not be ascribed to anything at all?[26]

---

[22] Ibid., page 50.

[23] Ibid., page 51.

[24] Ibid.

[25] *Proceedings of the Society for Psychical Research*, Volume 2, 1884, page 191. http://www.lexscien.org/lexscien

[26] Immanuel Kant, trans and ed P Guyer and A W Wood, *Critique of Pure Reason* (Cambridge: CUP, 1998), page 157.

In Lodge's increasingly imaginative model, the universe might possess an independent existence *to a degree*, but that existence was intimately bound to human consciousness through a skein of complex etheric interactions. In his 1912 work *Modern Problems*, Lodge would assert that space and time were not mere simplifications or reductions of the universe by the mind: 'They are not *directly* apprehended but they *are* apprehended, and they are real'.[27] Consciousness was therefore a manifestation of the ether and a thread of an unfolding universe that had both direction and purpose. Allied to the dynamic concept of mental evolution embraced by Alfred Russel Wallace, this new model of potentially unified forces was extremely persuasive to those seeking to preserve the spiritual in an increasingly materialistic version of the universe. Years later, in his hugely ambitious work *Making of Man*, Lodge fully linked his theory of ether to a concept of progressive evolution: 'An eye is an organ, not even yet completely understood, whereby the organism entered into relation with the Ether of Space and made use of its vibrations'.[28] Lodge's ideas were part of a complex philosophy by which living creatures were driven to perceive and understand the *ultimate* conduit of forces: the ether.

As we have seen, in the centuries since the overturning of the Ptolemaic model, humanity had drifted from the centre of cosmic importance to its periphery. But Lodge proposed to triumphantly return us to the centre. Over time, his system came increasingly to look like a religious one: vast and all-encompassing.[29] It was this ambitious and unashamedly metaphysical project that would lead to his alienation from the SPR in the following years. Yet, as Lodge proved with his demonstration of his version of the transmitter, he was very much his own man. He was as much a philosopher—perhaps even a theologian—as a scientist. He was a system builder and, by the 1880s, he believed he was close to discovering what we might term a unified 'theory of everything'. But the key component of that system, indeed its very adhesive, was about to be called into question.

---

[27] Oliver Lodge, *Modern Problems, The Nature of Time*, (London: Methuen, 1912), page 15.

[28] Oliver Lodge, *Making of Man: A Study in Evolution*, (London: Hodder and Stoughton, 1924) page 75.

[29] *Making of Man* frequently strikes a mystical tone at odds with much of Lodge's earlier work.

CHAPTER 10

Ether/Or…

Abstract In the final chapter, I examine the search for the 'missing' component in the Victorian view of the universe, the ether of space. Lodge believed it existed and enabled interactions between mind and matter, but when experiments failed to confirm its existence, he set about a range of new investigations with mediums, including Eusapia Palladino. These new investigations appeared to demonstrate a range of psychic and physical interactions, but distanced Lodge's research from the emerging scientific mainstream.

Keywords Ether • Space • Séance • Telekinesis • Ectoplasm • Psychic force • Oliver Lodge • SPR • Arthur Conan Doyle • Michelson and Morley • Eusapia Palladino

> While it is never safe to affirm that the future of Physical Science has no marvels in store even more astonishing than those of the past, it seems probable that most of the grand underlying principles have been firmly established (Albert Michelson, *Annual Register of The University of Chicago*, 1894).[1]

---

[1] Qtd in the *Annual Register*, July 1895–July 1896, page 159. Michelson was Head Professor of the Department of Physics.

Between April and July 1887, in the basement of a dormitory at the Case School of Applied Science in Cleveland, Ohio, Albert Michelson and Edward Morley tested their new machine: the interferometer. It required stillness and silence, since it was designed to detect the presence of that most elusive yet ubiquitous of things: the ether. In fact, it was designed to confirm its existence—and to consolidate the huge gains made in physics since Newton's *Principia*. Proof of the ether was to be the capstone of the project of understanding the physical universe. The hypothesis to be tested was this: light appears to travel in waves from the sun to the earth. All waves require a medium to move through. This substance might constitute the material with which supposed 'empty space' was filled: the ether. Perhaps light 'vibrates' within the ether and, if so, could it be detected? If the ether were a static phenomenon, then how might it respond to the movement of the earth? As John Gribbin has written:

> The experiment would involve splitting a beam of light in two and sending each of the two resulting beams on a journey between two mirrors, one set of mirrors lined up in the direction of the Earth's motion through space (and, presumably, through the ether) and the other at right angles to it. After bouncing between their respective mirrors, the beams of light could be brought back together and allowed to interfere. If the experiment was set up so that each beam of light covered the same distance, then because of the Earth's motion through the ether they should take different times to complete their journeys, and get out of step with one another, producing an interference pattern like the one seen in the double-slit experiment.[2]

But the experiment found no evidence to suggest the presence of the ether and, crucially, the speed of light was recorded as being exactly the same in every iteration of the test conducted between spring and summer of 1887. It appeared that the ether, if it existed at all, was deeply resistant to investigation.

Oliver Lodge, always magnanimous in matters of academic rivalry, was nonetheless unwilling to give up on the ether and attempted his own experiment. In 1893, he constructed an 'Ether Whirling Machine' in his laboratory in Liverpool. In an image perhaps subconsciously inspired by the work he'd observed in the potteries as a child, Lodge commissioned:

---

[2] John Gribbin, *Science: A History*, (London, Penguin, 2003) pages 436–437. The experiment, which sought to demonstrate the wave pattern of light, was first undertaken by Thomas Young in 1801.

a pair of discs, a metre in diameter, made at Sheffield, guaranteed to stand sixty tons to the square inch, and then asked Mather & Platt to make a frame or stand which would hold these discs mounted so as to be driven by one of their compact 'Manchester' dynamos, arranged on a vertical axis, with the shaft of the whirling discs in line with the axis of the dynamo, and spun by it so the discs spun horizontally...I also prepared a stone altar on which to mount the machine.[3]

The discs spun at 4000 RPM and for the optical part Lodge used the same set-up as Michelson and Morley had used in Cleveland, with silvered mirrors arranged so that 'the light travelled round and round in the space between the steel discs'.[4] By rapidly moving matter, the experiment attempted to demonstrate a measurable drag of light on account of the presence of the ether. But the experiment again met with disappointing results. Years later, Lodge reflected on this series of events. Despite adding a third disc and experimenting with a large piece of soft iron:

> The result of all the experiments was that nothing that I could do would alter the velocity of light one iota. We had no 'theory of relativity' then; but now it is a postulate of that theory that the velocity of light in space is an absolute constant.[5]

The failure of the 1893 experiment must have been a crushing blow to Lodge. The idea of ether allowed his imaginative speculation about the interaction of mind with matter to remain relevant. He received some comfort from the Irish physicist George Fitzgerald who came up with a plausible explanation of the failure to detect the movement of the ether in both his experiment and Michelson and Morley's: a shortening in the direction of motion that would create a level of distortion that might conceal the ether's effect. Lodge wrote excitedly in *Past Years*:

> The whole hypothesis was immediately afterwards illuminated and consolidated by (Hendrik) Lorentz, who showed that if the particles of all solids were held together by electrical forces...an effect of this kind was bound to occur; for motion of the body as a whole would superpose upon the electrostatic forces between the atoms a magnetic effect due to their motion. And

---

[3] Oliver Lodge, *Past Years*, (London, Hodder and Stoughton, 1931), pages 195–196.
[4] Ibid., page 197.
[5] Ibid., page 199.

on working this out he found that the amount of shrinkage was exactly what was needed to neutralize the effect otherwise to be expected in the Michelson experiment, and to give the result zero, as was actually obtained.[6]

Concluding a chapter of his autobiography on his scientific work in Liverpool, Lodge wrote of the ether: 'Let us try to keep the question open'.[7] Ultimately, the ether might serve as a theoretical bridge that could make sense of everything. Just as string theory in the twentieth century joined the worlds of quantum and super-massive space into a comforting continuum, so the ether permitted Victorian scientists to keep the notion of a purposeful intelligence at the heart of a universe of seemingly boundless and unfathomable powers.

But what did the continuing belief in the ether mean to Lodge personally? It supplied a space in which he could reconcile the extraordinary phenomena that he had witnessed with his own scientific training; a place in which he could imagine a unifying power and a way of filling the vacuum of space with something meaningful and, perhaps, knowable. The idea of the ether was not abandoned, and Lodge rarely allowed himself to become downcast. Indeed, in an intellectual counter-offensive against recent disappointments, his experiments in the area of mental power began to take on a new intensity.

In 1889, Lodge had met the medium Leonora Piper when she arrived at Liverpool from the United States. Piper had given a number of séances in Boston that had convinced William James—Professor of Psychology at Harvard and brother of the author Henry James—that she possessed genuine powers of telepathy. This success led to the SPR inviting Piper to Britain to undertake a series of tests. The initial tests were conducted by Frederic Myers in Cambridge, but Lodge soon joined the investigation. The séances Piper, Myers and Lodge shared would have a profound impact on the physicist's world view:

> The result was quite astonishing. Messages were received from many subordinate people, but the special feature was that my aunt Anne, who has so often been mentioned before in connection with my education, and who was now on the other side – having succumbed to an operation for cancer – ostensibly took possession of the medium; and in her own energetic manner reminded me of her promise to come back if she could, and spoke a few

[6] Ibid., page 206.
[7] Ibid., page 207.

sentences in her own well-remembered voice. This was an unusual thing to happen, but was very characteristic of her energy and determination. The sitting continued till midnight, a great deal more was said, and Myers and I were both exhausted when it was over.[8]

Having experienced what he considered to be a genuine and personal communication from his beloved aunt—the person who had inspired his scientific journey—via Leonora Piper, Lodge invited the medium to participate in further tests. Piper communicated additional messages from his relatives, before apparently demonstrating an astonishing phenomenon:

> the reading of an unopened letter applied to the top of her head, a phenomenon which had already been testified to by Kant and Hegel, though by them it was called 'reading with the pit of the stomach'. At any rate, it was reading without the use of the sense organs, and therefore represented another obscure human faculty commonly called 'clairvoyance'.[9]

It's hard to know what to make of Lodge's meetings with Leonora Piper. They seem as extraordinary as Crookes' encounters with D D Home 18 years earlier. The results that Lodge first experienced in his early investigation of telepathy in Liverpool undoubtedly laid the foundation for his experiences with Piper, but he was clearly unprepared for the personal and emotional nature of the communications. In *Past Years*, he describes how a Cambridge graduate called Richard Hodgson became so obsessed with Piper that he hired a detective to shadow her movements in Boston, hoping to discover a visit to a library, public records office or even a cemetery. But no such evidence presented itself and 'Hodgson gradually satisfied himself that she did none of these things, but lived a perfectly normal life with her children'.[10] For Lodge, the only conclusion was to accept Piper as a genuine medium, a telepath and—possibly—a person in communion with the dead. Despite the criticisms of such phenomena by Crookes' old rival W B Carpenter and his description of cold reading techniques and 'fishing', Lodge's belief in the integrity of Leonora Piper remained. Lodge's response was to return again to the ether. At the same time as he was developing a new machine for detecting electro-magnetic waves—the 'coherer'—Lodge mapped the possibility of mental communication into

[8] Ibid., page 277.
[9] Ibid., page 278.
[10] Ibid., page 282.

an even more elaborate model. The ether was not only the substance that filled the universe and perhaps even unified it, but it also interacted with and stored the product of human thought as *memory*. In the next phase of his research, Lodge encountered an Italian medium who further convinced him of the ether's transcendent qualities: Eusapia Palladino.

Palladino was born in Bari, Italy, in 1854. She was an orphan who lived in poverty until she was adopted by a family in Naples. Palladino received no formal education and spoke only the local Italian dialect she'd grown up with, even when she became internationally famous. Interestingly, her first marriage was to a magician, Raphael Delgaiz, and she assisted him on tour and in the running of his shop, a fact Lodge seemed to be unaware of, claiming in his autobiography that she sold baby linens 'on the Bay of Naples'.[11] From the late 1880s, Palladino conducted a number of séances and demonstrated a range of phenomena, including levitations, spirit lights, rapping and table-turning. As in the case of the Fox Sisters, these activities led to invitations to perform across Europe and gained the attention of the French physiologist Charles Richet—a friend of Lodge—who undertook a series of test séances with her, inviting members of the British SPR to his home on the Ile Roubaud in the Mediterranean in summer, 1894, to see the phenomena for themselves. One year after the failure of his ether-detecting machine, Lodge set out to meet one of the most controversial mediums since D D Home.

In *Past Years*, Lodge paints an idyllic picture of life on the island. Séances were held in the evening in conditions that clearly lacked the restrictions of Crookes' Mornington Road experiments with Home, or the electrical tests of Cromwell Varley:

> Two of our number, sometimes Richet and Myers, sometimes Myers and I, sat on either side of Eusapia, holding each one hand and constantly ejaculating for the benefit of the others, whenever anything occurred, 'J'ai la main gauche', 'J'ai la main droite', making sure which hand it was by the position of the thumb.[12]

Lodge acknowledged that Palladino was capable of deception, which she explained away as the influence of her mischievous 'control', but he continued to believe that the majority of her phenomena were genuine. At

---

[11] Ibid., page 296.
[12] Ibid., page 295.

one point, Richet placed an instrument called a dynamometer in the room. It measured grip strength and at one point, Palladino's recorded an extraordinarily high level: 'to our surprise we saw the needle going up, indicating a force far beyond what any of us could exert'.[13] Alongside this, Palladino generated exceptionally loud rappings that 'sounded like blows delivered with a heavy mallet',[14] and, like Home, caused instruments to play and an unwound mechanical music box to work by itself.

However, the most astonishing phenomena were yet to come. During the final sitting on Ile Roubaud, Palladino went into a trance and appeared to demonstrate the most direct and powerful evidence of force acting upon an object:

> There was an escritoire standing against the wall, about a yard from her, and she proceeded to make a gesture towards it, as if she were intending to reach it or push it, her hands being well and visibly held. Every time she did this, the piece of furniture tilted back against the wall, just as if she had had a stick in her hand and was pushing it. This happened three times.[15]

Lodge used the term that Richet had coined to describe this force: ectoplasm. The movement of furniture in the room was not caused by 'psychic force', but by some reaction between the medium's body and the air around the medium and sitters that allowed for the realization of 'a sort of supernumerary arm, which occasionally I could see, once sitting behind her for the purpose, while Myers and Richet were holding her normal arms'.[16] Richet was baffled by what took place in the room but concluded (according to Lodge): 'C'est absolument absurd, mais c'est vrai'.[17] Reflecting on his encounter with Palladino in the *Journal of the SPR* in November 1894, Lodge wrote the following:

> The phenomena do not seem to modify the fundamental laws of physics, but perhaps they may lead to an extension of the recognised laws of biology. In other words, it is only in the presence of a living being that these actions occur, and the power which enables such movements appears to be a modified or unusual display of vital power, directing energy in an unusual way

---

[13] Ibid., page 298.
[14] Ibid.
[15] Ibid., page 301.
[16] Ibid., page 301.
[17] Ibid., page 302.

along unrecognised channels…instead of action at a distance in the physical sense, what I have observed may be said to be like vitality at a distance.[18]

Although Lodge, Myers and Richet were convinced by Palladino's apparent telekinetic abilities, she failed to convince the Sidgwicks. An accomplished linguist as well as a leading mathematician, Eleanor had served as interpreter between the various parties in the séance. She saw nothing in the séance room at Ile Roubaud to convince her. During further investigations in Cambridge, the SPR claimed to have caught the medium in multiple instances of fraud. Lodge defended her using the same arguments as before—the 'control' was mischievous, the force at work was inconsistent, and it manifested itself in a variety of different ways.

Fourteen years after the founding of the SPR, the fault line between practical scientists like Lodge and the investigators from the social elites, such as Eleanor and Henry Sidgwick, had not closed up. Decades later, Arthur Conan Doyle would dedicate his *History of Spiritualism* to Lodge and make his differences with the SPR public. Both Lodge and Doyle lost sons in World War I and invested more and more emotional and intellectual energy in attempting to prove that personal consciousness might survive death. What was most remarkable about Oliver Lodge's relationship with mysterious forces was the fact that it wasn't an interest he stumbled upon later in his life, it was *always* a preoccupation and he took part in séances at the same time as his investigation of electro-magnetism and radio waves. His significance as a scientist may have been reduced on account of his connection to spiritualism, but it was as much a part of his motivation and intellectual process as alchemy was to Newton's. Lodge never lost the Kantian impulse to attempt to build a theory of everything and to retain the importance of humanity as an important agent at the heart of a baffling, but ultimately explicable, cosmos. If this led to credulousness—and it's hard to defend the practice of theatrical mediums such as Palladino or to deny her associations with magicians—then it is forgivable in a mind that considered no phenomenon beneath investigation, one that consistently sought to make the purpose of science accessible and meaningful to working people.

To Lodge and many of his contemporaries, the identification of the ether was tantalizingly close. It had stubbornly refused to reveal itself to Michelson, Morley and Lodge himself, but the fact that light *must* travel

---

[18] Ibid., page 307.

through some kind of medium remained a reasonable hypothesis up to Einstein's theory of General Relativity. Indeed, Einstein did not believe that he had disproved the ether by establishing the existence of the photon in 1905. The question of what space is made from remains with us today, and vast regions of the universe are given the mysterious name 'Dark Matter'. What Lodge sought to do, possibly unconsciously, was to create a field of speculation that was spiritual in nature and gave a sense of grand purpose. Lodge physicalized the ether into a force that could, potentially, be interacted with. In his 1924 book *Making of Man*, dedicated to Frederic Myers, Lodge wrote: 'The association of Spirit with Matter, the incarnation of something pre-existent, is a reality, whether we understand it or not'.[19] Did he mean ether was the substance of, or a conduit for, the soul? He was certainly attempting to create a system that reconciled faith with science. This, to many of his contemporaries and perhaps to us, was simply folly. Yet, as a proposal designed to affirm the significance of human action, it was both uplifting and inspiring. Lodge's mission was to attempt to understand and explain the forces that shape the universe to anyone who would listen, and to express this with a progressive vision that emphasized its beauty and unfolding mystery. As he said in his lecture 'Modern Views on Matter', given at The Sheldonian Theatre in Oxford in the summer of 1903:

> I believe that the simplicity and beauty of the truth concerning even the material universe, when we know it, will be such as to elicit feelings of reverent awe and adoration.[20]

The sentiment was equivalent to a manifesto that he never reneged upon. Lodge possessed a brilliant mind and never worked for material benefit. He was an evangelist for science, in the tradition of Michael Faraday, who sought to invite his audience to experience a process that was, almost magically, revealing itself to them. It was illuminated via the work of investigators who were the sons of potters, blacksmiths and tailors. It was as much a chivalric *quest* as an exploration. In the twenty-first century, as

---

[19] Oliver Lodge, *Making of Man*, page 167.
[20] Oliver Lodge, *The Romanes Lecture: Modern Views on Matter*, (Oxford: OUP, 1907), page 24.

more and more extraordinary and exotic phenomena are discovered in modern physics, the need for generous and inspirational interpreters is increasingly apparent. We really should know a great deal more about Oliver Lodge.

CHAPTER 11

Epilogue: Even Stranger Things

Abstract In the epilogue, I draw together some of the themes that have developed in this book. I consider how the concept of the ether was never fully abandoned by Lodge or by Albert Einstein and speculate—fancifully—on the nature of the quantum world. I use the idea of the Implicate Order, derived from the work of the American scientist David Bohm, to demonstrate our continuing desire to see new iterations of the mysterious and the occult in the world around us.

Keywords Relativity • Quantum • Implicate Order • God • Einstein • Spooky action • David Bohm

> The most beautiful thing one can experience is the mysterious. It is the source of all true art and science (Albert Einstein, *My Credo*).[1]

My objective in this short book has been to present a brief history of mysterious and occult forces from Newton to the Victorians, but also to consider how people reacted to such forces: endorsed them, resisted them,

---

[1] Albert Einstein, trans. H J Kupper, *For the League of German Human Rights*, c. 1932, https://einstein-website/de/en/credo

feared them or embraced them. It's been, I hope, an interesting journey and I'm not keen to see it end. As such, I can't resist considering what came next in the 'after life' (so to speak!) of these forces and sharing some, frankly, rather speculative final thoughts.

The year 1905 saw the sidelining of the theory of the ether, with Albert Einstein's Special Theory of Relativity describing the action of the photon. This demonstrated that light travels through the vacuum of space without a wave-bearing medium. There was, supposedly, no further need for the ether. As with so many of the great minds presented here, Einstein was not part of the social or intellectual elites of Europe. He was Jewish and encountered antisemitic prejudice throughout his career. But it's likely the freedom of being an outsider allowed him to think beyond the constraints of academic physics and to pursue his broader interest in philosophy. In 1905, he was working as a patent clerk having failed to achieve even an entry-level job in Swiss academia. As Walter Isaacson has written: 'Had he given up theoretical physics at that point, the scientific community would not have noticed'.[2] But he was drawn to the strange nature of the Planck Constant, a sum that equalizes the energy of an electromagnetic wave with its frequency. Einstein applied this to the problem of light with results that changed the study of physics forever. Light travelled not in waves but in 'quanta', packets of energy that had the qualities of both wave and particle. Einstein's friend and colleague, Niels Bohr, applied the new model to the hydrogen atom, revealing once again the existence of wave/particle duality. Just a few decades before, the Victorians believed that they had discovered the key elements at work in the universe, with only the ether left to confirm. But these new discoveries would lead to an entirely new theoretical model: quantum mechanics. The primacy of the speed of light was confirmed by Einstein and, in the General Theory of Relativity ten years later, light was shown to bend in the face of gravity, but never to slow down. Light speed was unchanging, regardless of the position or frame of reference of the perceiver. As the subatomic world revealed itself over time, the laws of classical physics were capsized by the actions of particles that simply failed to conform to them. The division of the atom would have profound practical applications in terms of energy and, terrifyingly, weaponry. By the middle of the twentieth century, humans understood the potential of mysterious forces so well that they could use them to destroy their planet thousands of times over. The destructive potential

---

[2] Walter Isaacson, *Einstein*, pages 92–3.

of atomic power, the idea of which prompted Robert Oppenheimer to quote the *Bhagavad Gita*: Chapter Eleven, Verse Thirty-two: 'Now I become Death, the Destroyer of Worlds'[3] continues to represent one of the greatest challenges to humanity in the twenty-first century.

Since Schopenhauer, some western philosophers and scientists have seen a version of reality in ancient Indian texts that seemed more credible than the one they had inherited from their own faith traditions. Supposedly, Einstein had four portraits in his study during his time at Princeton: Schopenhauer, Faraday, Maxwell Clerk and Gandhi. The concept of an underlying reality, some unifying force beneath the world we perceive, was deeply important to him. As he was reported to have said in a conversation with Harry Kessler:

> Try and penetrate with our limited means the secrets of nature and you will find that, behind all the discernible laws and connections, there remains something subtle, intangible and inexplicable.[4]

Even when he had removed the *reason* for a theory of the ether, he was not keen to abandon the idea entirely. In a lecture given at the University of Leyden in 1920, Einstein stated that the idea of the vacuum of space possessing no physical qualities was probably incorrect:

> there is a weighty argument to be adduced in favour of the ether hypothesis. To deny the ether is ultimately to assume that empty space has no physical qualities whatever. The fundamental facts of mechanics do not harmonize with this view.[5]

The classical concept of space having substance and structure had not entirely gone away. But, as the study of quantum mechanics advanced, so the possibility of a random and unpredictable universe grew. The position of Bohr, Werner Heisenberg and their allies—known as the Copenhagen Interpretation—offered little comfort to those seeking evidence of some all-shaping intelligence in the universe. Einstein was seeking a theory of

---

[3] There are, of course, multiple English translations of this ancient text. Here is a more contemporary version by Dr. Ramananda Prasad, https://www.gita-society.com/gita-in-english-chapter11

[4] Albert Einstein to Harry Kessler, qtd in Isaacson page 384.

[5] Albert Einstein, *Lecture on Ether and the Theory of Relativity,* University of Leyden, 5th May 1920, https://www.gutenberg.org/files/7333/7333

everything, with special relativity as the first step, yet the emerging world of quantum was random and apparently directionless. Classical concepts of order broke down on the subatomic level. The ensuing crisis inspired one of Einstein's most famous statements in a letter to Max Born: 'In any event, I am convinced that He (God) is not playing dice'.[6] And yet He certainly *appeared* to be.

In one of the most baffling examples of activity in the quantum world, two particles can be set to resonate at the same frequency and then be sent far away from each other. But whatever is done to the first particle manifests in the second regardless of the distance. They have become 'entangled'. Einstein, whose theory identified light as the fastest thing in the universe, dismissively labelled this activity 'spooky action at a distance'.[7] Yet, as hard as he tried, he could not fully account for it. If a force is being exercised upon one particle by the other, then it is apparently travelling *faster* than light. To explain this, a range of speculative solutions have been offered with none, so far, proving to be entirely satisfactory. As with the early uses of electric current, our ignorance has not prevented us from making ample use of 'spooky action' in modern communications technology. Despite our 'domestication' of this phenomenon, quantum mechanics presents us with the most astonishing mystery so far. Einstein did not foresee the manifold, exotic complications, and he was always sceptical about the precise nature of spooky action.

Albert Einstein continued in his quest for unity. But is there something at work that we haven't yet understood? If so, is it *possible* for us to know it? Have we wrongly conditioned ourselves to seek unity? Perhaps it will take another revolution of viewpoint for us to make such a breakthrough. In 1952, an assistant of Oppenheimer called David Bohm offered just such a vision. Bohm presented space not as a vacuum through which things travel, but as an active, informing entity that gives particles their mass and purpose. It is, therefore, in continuous dialogue with the consciousness of everything that perceives it. Waves, particles and quanta are phenomena, part of a far greater fabric: the Implicate Order. What we are experiencing is the unfolding of this order and, in this unravelling from implicate to explicate, new characteristics become known to us. Perhaps Oliver Lodge would have approved.

---

[6] Albert Einstein, *Letter to Max Born*, 4th December 1926. Doc. 426, https://einsteinpapers.press.princeton.edu

[7] Walter Isaacson, page 232.

Bohm, the son of a Jewish furniture salesman from Wilkes Barre, Pennsylvania, frequently described his version of the quantum universe in terms associated with Vedic philosophy. His work is disputed by physicists with many regarding it as fanciful. And yet it has that quality of awe and mystery that we remove from science (and from life) at our peril. Gray, Faraday, Crookes and Lodge were, at least initially, outsiders to the scientific establishment whose work was inspired by the skilled manual labour of their ancestors and a sense of purpose they inherited from religious faith. Often, their discoveries challenged and even overturned that faith, but what matters most is the sense of a grand narrative that these scientists possessed. Even if we reject Bohm's theories, we are still faced with a world of quanta that are bound up with our own consciousness. Particles exist in several states—'superpositioned'—apparently 'choosing' one in order to be perceived by us. But why do they do it? Why waste such colossal amounts of energy in the process? Why expend so much effort merely to be *apprehended* by a gang of big-headed primates on a spinning rock? Science can reveal some of the chain of cause and effect to us, but it can't tell us *why* it's happening. Yet we seem cursed, or perhaps blessed, to keep asking that question.

I began this book with an observation about Isaac Newton. When he became Lucasian Professor of Mathematics at Cambridge, he spent much of his stipend on alchemical texts and a pair of ovens for melting metals. In the last three centuries or so, we have turned Newton into the respectable poster child for 'legitimate', laboratory-based science, but his motivation was the greatest of all: to gain a greater understanding of the mysteries around him. At heart, he remained an alchemist. I'm not suggesting that we do the same, but I will assert the importance of the human narrative in the approach to science. Science and philosophy, indeed science and the humanities, should always walk together, towards whatever new mysteries experimentation suggests. There may be no noumenon, no philosopher's stone, no implicate order and no creator. But we'll never know unless we look.

Bibliography

Ackroyd, P., *Introducing Swedenborg* (London: The Swedenborg Society, 2021).
Aldini, J., *An Account of the Late Improvements in Galvanism with a Series of Curious and Interesting Experiments* (London: Cuthell and Martin, 1803).
Aristotle, trans. R P Hardie and R K Gaye *The Physics. Writings on Natural Philosophy* with a New Introduction by Dr James Lees (London: Flame Tree 2023).
Braid, J., *Neurypnology* (London: George Redway, 1899).
Burns, J., (ed.) *Report on Spiritualism of the Committee of the London Dialectical Society Together with the Evidence Oral and Written and a Selection from the Correspondence* (London: Longman, Green, Reader and Dyer, 1871).
Cartwright, D., *Schopenhauer: A Biography* (Cambridge: CUP, 2013).
Clarke, A.C., *2001: A Space Odyssey: 50th Anniversary Edition* (London: Orbit, 2018).
Clodd, E., *The Question: A Brief History and Examination of Modern Spiritualism*, (London: Grant Richards, 1917).
Conan Doyle, A., *The History of Spiritualism*, Vol. 2, (London: Cassell and Company, 1926).
Copp, Rev. J.A., *The Atlantic Telegraph: As Illustrating the Providence and Benevolent Designs of God. A Discourse, Preached in the Broadway Church, Chelsea, August 8, 1858.* (Boston: T R Marvin and Son, 1858).
Crookes, W., *Researches in the Phenomena of Spiritualism Reprinted from the Quarterly Journal of Science*, (London: J Burns, 1874).
De la Mettrie, J.O., trans and notes G C Bussey, *Man a Machine* (Chicago: The Open Court Publishing Co., 1912).

Dunraven, William Windham Thomas Quinn, Lord Adare, *Experiences in Spiritualism with Mr D D Home* (New York: Arno Books, 1976).

Elliotson, J., *The Harveian Oration* (London: H Bailliere, 1846).

———. Human Physiology (London: Longman, Rees, Orme, Brown, Green and Longman, 1835).

——— *Numerous Cases of Surgical Operations Without Pain in The Mesmeric State* (Philadelphia: Lea and Blanchard, 1843).

Faraday, M., *The Chemical History of a Candle* (Cromwell Collier, 1962).

Fournier D'Albe, E.E., *The Life of Sir William Crookes*, (London: T. Fisher Unwin Ltd, 1923).

Franklin, B. *The Autobiography of Benjamin Franklin* (Philadelphia: Henry Altemus, 1899).

Gribbin, J. Science: A History (London: Penguin 2003).

Gurney, E., Frederic W. H. Myers and Frank Podmore, *Phantasms of the Living* Vol. 1 (London: Trubner and Co.).

Hall, T.H. *The Medium and the Scientist: The Story of Florence Cook and William Crookes* (New York: Prometheus Books, 1984).

Hanegraaf, W.J., *Western Esotericism A Guide for the Perplexed* (London: Bloomsbury, 2013).

Hardinge, E., *Address Delivered at the Winter Soirees* (London: Thomas Scott, 1866).

Hardinge, E. *Modern American Spiritualism* (New York: The Author, 1870).

Hume, D., *An Enquiry Concerning Human Understanding* (Chicago: The Open Court Publishing Co., 1900).

Issacson, W., *Einstein: His Life and Universe* (London: Simon and Schuster, 2007).

Kant, I., trans and ed P Guyer and A W Wood, *Critique of Pure Reason* (Cambridge: CUP, 1998).

———. trans. E F Goerwitz and Intro by F Sewall, *Dreams of a Spirit Seer Illustrated by Dreams of Metaphysics* (London: Sonnenschein and Co., 1900).

Kerner, J., trans. A M H Watts and Preface by C Mingins. *Franz Anton Mesmer, the Discoverer of Animal Magnetism* (Hidden Tarn Editions, 2020).

King, W.D., *Henry Irving's Waterloo: Theatrical Engagements with Arthur Conan Doyle, George Bernard Shaw, Ellen Terry, Edward Gordon Craig, Late-Victorian Culture, Assorted Ghosts, Old Men, War and History* (Berkeley: UCP, 1993).

Lachman, G., *Introducing Swedenborg: Correspondences* (London: The Swedenborg Society, 2021).

Lamont, P., *The First Psychic: The Peculiar Mystery of a Notorious Victorian Wizard* (London: Abacus, 2006).

Lewis, L. *Henry Irving and The Bells: Irving's Personal Script of the Play*, ed. D. Mayer (Manchester: MUP 1980).

Lodge, O., *Ether and Reality: A Series of Discourses on the Many Functions of the Ether of Space*, (Cambridge: CUP Library Collection, 2012).

Lodge, O. *Modern Problems*, The Nature of Time, (London: Methuen, 1912).
———. *Modern Problems Past Years: An Autobiography*, (London: Hodder and Stoughton, 1931).
——— *The Romanes Lecture: Modern Views on Matter*, (Oxford: OUP, 1907).
McClenachan, C.T., *Detailed Report of the Proceedings Had In Commemoration Of The Successful Laying Of The Atlantic Telegraph Cable* (New York: E. Jones, 1863).
Magee, B., *The Philosophy of Schopenhauer* (Oxford: OUP, 1983).
McGarry, M. Ghosts of Futures Past (Berkeley and Los Angeles, 2008).
Melechi, A., *Servants of the Supernatural: The Night Side of the Victorian Mind*, (London: William Heinemann, 2008).
Moore, W., *The Mesmerist: The Society Doctor Who Held London Spellbound* (London: Weidenfeld and Nicolson, 2018).
Morus, I.R., *Shocking Bodies: Life, Death and Electricity in Victorian England* (Stroud: The History Press, 2011).
Newton I., trans. A. Motte, *Newton's Principia The Mathematical Principles of Philosophy* (New York: Daniel Ader, 1846).
Noakes, R. *Physics and Psychics The Occult and the Sciences in Modern Britain* (Cambridge: CUP, 2019).
Noakes, R. *Physics* 'Spiritualism, science and the supernatural in mid-Victorian Britain' *The Victorian Supernatural*, eds. Bown, Burdett and Thurschwell (Cambridge: CUP, 2004).
Oppenheim, J., The Other World: Spiritualism and Psychical Research in England, 1850-1914 (Cambridge: CUP, 1985).
Roberts, A., 'Browning, reincarnation and the resuscitation of the dead' in *The Victorian Supernatural*, eds. Bown, Burdett and Thurschwell (Cambridge: CUP, 2004).
Rovelli, C., *Anaximander and the Nature of Science* (London: Allen Lane, 2023).
Russell, W.H., *The Atlantic Telegraph* (Alpha Editions, 2022).
Russel Wallace, A., *Miracles and Modern Spiritualism* (London: George Redway, 1896).
Russel Wallace, A. *My Life A Record of Events and Opinions* (London: Chapman and Hall, 1905).
Schopenhauer, A., trans. E F J Payne, *The World as Will and Representation*, Volume One (New York: Dover, 1969).
——— trans. J. Frauenstadt, *Two Essays by Arthur Schopenhauer* (London: George Bell and Sons).
Shelley, M., *Frankenstein or The Modern Prometheus* (Oxford: OUP, 1984).
Sidgwick, E., *Mrs Henry Sidgwick: A memoir by her niece*, (London: Sidgwick and Jackson, 1938).
Slotten, R.A., *The Heretic in Darwin's Court* (New York: Columbia University Press, 2004).

Stukeley, W., *Memoirs of Sir Isaac Newton's Life* (London: Royal Society, 1752).
Swedenborg, E. trans. C TH Odhner, *On Tremulation* (Boston: MNCU, 1899).
────── trans. J. Randel and E. Tansley, *The Principia or the First Principles of Natural Things* (London: The Swedenborg Society, 1912).
Von Reichenbach, K., *The Odic Force or Letters on Od and Magnetism*, trans. F D O'Byrne (University Books, 1968).
Waddell, M., *Magic Science and Religion in Early Modern Europe* (Cambridge: CUP, 2021).
Weisberg, B., *Talking to the Dead: Kate and Maggie Fox and the Rise of Spiritualism* (New York: Harper Collins, 2004).
White, M., *Isaac Newton The Last Sorcerer* (London: Fourth Estate, 1997).

Periodicals

The Daily Telegraph
The Lancet
The Medium and Daybreak: A Weekly Journal Dedicated to the History, Phenomena, Philosophy and Teachings of Spiritualism
The Spiritualist
The Zoist

Online Resources

https://www.biblegateway.com
https://dictionary.cambridge.org
https://einsteinpapers.press.princeton.edu
https://einstein-website./de/en
https://www.gita-society.com/gita-in-english
https://www.gutenberg.org
http://iapsop.com/archive/materials
https://www.lexscien.org/lexscien
https://www.spr.ac.uk
https://westminster-abbey.org

Author Index[1]

A
Adare, Lord (W.T. Wyndham Quinn, 4th Earl of Dunraven), 82
Aldini, G., 53, 54, 54n11, 57
Anaximander, 9, 14

B
Balfour, A., 108
Barrett Browning, E., 82
Barrett, W.F., 109–111
Beckett, S., 27
Blackburn, C., 95
Bohm, D., 130, 131
Born, M., 130
Bose, M., 50, 52
Braid, J., 40–42, 44, 45
Britten, E.H., 82, 92, 93
Browning, R., 82, 83, 85
Buchanan, J., 60
Bulwer-Lytton, E., 82
Burns, J., 81n6

C
Carpenter, W.B., 101–103, 108, 121
Charles II, King of England, 8
Charles XII, King of Sweden, 21
Conan Doyle, A., 109, 124
Cook, F. ('Katie King'), 89, 91–101, 109
Cook, K.S., 92
Copp, Rev. J., 60, 61
Cox, E., 84, 85, 87, 95, 97, 102
Creery, Rev. T. (and daughters), 111–114
Crookes, P., 83, 99
Crookes, W., 3, 68, 73, 80, 81, 83–91, 93n14, 95–103, 106, 109, 121, 122, 131

D
Darwin, C., 70–75
Davenport Bros, The, 97
De la Mettrie, J. O, 47, 55–57, 62

[1] Note: Page numbers followed by 'n' refer to notes.

Delgaiz, R., 122
Descartes, R., 22
Dickens, C., 39
DuPotet, Baron (Jules Dennis Dupotet Sennevoy), 34–36, 83

E
Einstein, A., 2, 14, 26, 78, 108, 125, 127–130
Elliotson, J., 14, 31–45, 33n4, 36n14, 70, 79, 87, 100, 103, 108, 110
Esdaile, J., 39

F
Fairlamb, A., 92
Faraday, M., 13, 49, 58, 59, 78, 80, 90, 99, 101, 106, 107, 125, 129, 131
Forster, G., 54
Fox, K., 81–83, 90, 94, 97, 122
Fox, M., 81–83, 90, 97, 122
Fox-Fish, L., 81–83, 97, 122
Franklin, B., 20, 48, 51–53, 59, 60, 62
Freud, S., 40

G
Galvani, L., 3, 52, 53, 57, 61, 66, 79
Gray, J., 94
Gray, S., 49, 96, 131
Gurney, E., 109–112, 114

H
Hauksbee, F., 49–51, 80
Hegel, G.W.F., 25, 27, 121
Heisenberg, W., 129
Herne, F., 95
Hertz, H., 106
Home, D.D., 81–88, 90, 91, 93, 95–97, 99–101, 121–123
Houdini, H., 97
Hume, D., 73

J
Jackson, C., 60
James, H., 120
James, W., 120
Jesus, 64

K
Kant, I., 14, 23, 25–27, 55, 56, 58, 75, 79, 107, 108, 114, 121

L
LaFontaine, C., 40
Lamont, P., 82n8
Livermore, C., 94
Lodge, O., 3, 73, 105–110, 113–115, 115n29, 118–126, 130, 131
Lorentz, H., 119
Luxmoore, J.C., 95

M
Marryat, F., 98
Martineau, H., 39
Maxwell, J.C., 77–79, 78n3, 88, 90, 106, 108, 129
Mesmer, F.A., 3, 14, 18–22, 24, 25, 28, 29, 31, 33, 34, 37, 40, 42, 50, 67, 68, 79, 83, 110
Michelson, A.A., 2, 4, 117–120, 117n1, 124
Morley, E.W., 2, 4, 118, 119, 124
Morse, S., 59–62
Myers, F., 108–112, 120–125

O
Okey, E., 35–37, 42, 43, 45, 108
Okey, J., 36, 37, 42, 43, 45
Oppenheim, J., 4
Oppenheimer, R., 129, 130
Owen, R., 71, 75

P
Palladino, E., 109n6, 122–124
Piper, L., 120, 121
Podmore, F., 110, 112
Puyser, Marquis de, 83

R
Rayleigh, Lord (J.W. Strutt), 109
Reichenbach, Karl von, 14, 65–70, 73–76, 80, 88, 90, 100, 114
Richet, C., 122–124
Rovelli, C., 9, 14

S
Schopenhauer, A., 25–29, 55, 66, 69, 75, 90, 129

Sidgwick, Edith, 92, 99
Sidgwick, Eleanor, 92, 109, 124
Sidgwick, H., 109, 124
Swedenborg, E., 14, 21–24, 29, 68

T
Thales, 9, 49, 77
Tiemann, D.F., 61

V
Varley, C., 79–81, 83, 84n13, 88, 90, 96, 97, 109, 122
Victoria, Queen of Great Britain, 60
Volta, A., 3, 52, 53, 57, 61, 66, 79

W
Wagner, R., 27
Wallace, A.R., 70–76, 73n17, 78, 79, 81, 84, 88, 108, 109, 115
Weisberg, B., 94
Wittgenstein, L., 26

Subject Index[1]

A
Alchemy, 8, 11, 18, 124
Animal magnetism, 14, 19, 20, 24, 28, 33, 38, 40, 43, 45, 68, 100
Arian doctrine, The, 8
Atlantic Telegraph, The, 60, 62, 78, 79, 106

B
Battery, 36, 39, 51, 55, 68, 74, 86, 96
Bhagavad Gita, The, 129

C
Cambridge, 11, 49, 50, 59, 108, 109n6, 110, 112, 120, 121, 124, 131
CERN, 2
Clairvoyance, 121
Conduits, 2, 49, 52, 57, 61, 65–76, 81, 87, 93, 106, 115, 125

D
Darwinism, 72

E
Ectoplasm, 123
Electricity, 3, 14, 22, 44, 48–55, 57–59, 61–64, 66–68, 74, 79, 80, 88, 93, 106
Empiricism, 25
ESP, 33
Ether, 2–4, 10, 12–14, 105–108, 110, 114, 115, 117–126, 128, 129
Evolution, 70, 72, 75, 115

F
Force, 2–5, 7–11, 14, 17–29, 31–45, 48–50, 52–55, 57, 58, 60, 64, 66–70, 73–88, 90, 99, 100, 102, 103, 106, 109, 114, 115, 119, 123–125, 127–130

[1] Note: Page numbers followed by 'n' refer to notes.

Frankenstein, 14, 44, 47–49, 53, 54, 56, 57, 64

G
Galvanism, 53, 54
Ghost, 68, 69, 87, 93, 100
God, 4, 8, 9, 11, 12, 22, 23, 53, 55, 56, 58, 60–62, 66, 90, 106, 108, 130
God Particle, The, 2
Gravity, 2, 7, 8, 10–12, 24, 28, 44, 64, 78, 106, 128

H
Hypnotism, 3, 29, 40, 41, 44, 45
Hysteria, 68

I
Implicate order, the, 130, 131

J
Jar Phoonk, 39

L
Large Hadron Collider, The, 2
Leyden Jar, The, 50, 53, 57, 62
London Dialectical Society, The, 72, 79, 81, 83, 84, 99, 100, 102, 108

M
Magic, 1, 3, 13, 29, 34, 49, 58, 59, 64, 94, 109, 111
Magnetism, 18, 29, 57, 58, 68, 88
Materialism, 4, 23, 57
Medium, 21, 65–76, 81–84, 90–102, 106, 109, 111, 114, 118, 120–125, 128
Melodrama, 3, 43
Mesmerism, 13, 20, 29, 32–42, 44, 67, 69, 79, 83, 84, 87, 102, 110
Mind, 2, 3, 10, 13, 14, 22, 29, 31–45, 55, 58, 64, 67, 72, 74–80, 88, 90, 100–102, 105–115, 119, 124, 125, 128
Mystery, 1, 2, 4, 10–14, 23, 25, 26, 29, 34, 42, 44, 49, 60, 76, 101, 125, 130, 131

N
Natural Selection, 70, 72, 74
Nature, 3, 4, 8, 9, 11–14, 18–20, 22, 24, 25, 27, 37, 41, 44, 45, 52, 55–57, 61–64, 66–70, 73–75, 80, 84, 85, 87, 88, 90, 99, 100, 112, 113, 121, 125, 128, 129
Nature Philosophy, 20, 27, 58, 67
Nerves, 21, 22, 38, 74
Neurypnology, 40, 44

O
Occult, 3, 5, 8, 11, 13, 40, 45, 50, 56, 57, 74, 88, 127
Odic force, 14, 67–69, 73, 74, 90, 100

P
Polar likeness, 68
Psychic force, 3, 77–88, 90, 100, 102, 103, 123

Q
Quakers, 71
Quantum mechanics, 2, 128–130

R
Relativity
  General Theory of, 26, 128
  Special Theory of, 14, 128

S
Sandemanians, 59, 80
Science, 3–5, 12, 14, 15, 29, 42, 49, 60, 62, 64, 66, 75–77, 81, 84, 87, 90, 101, 103n40, 124, 125, 127, 131
Séance, 19n5, 32, 73, 73n17, 74, 81–83, 90–101, 103, 108, 109, 120, 122, 124
Sensitive, 65–76, 80, 90, 110
Society for Psychical Research, 76, 101, 108
Somnambulism, 41
Spiritualism, 68, 70, 73, 76, 81, 82, 91–93, 97, 99, 101, 102, 103n40, 108, 109, 124
Spooky action, 130

T
Telekinesis, 13, 14, 23, 68, 87, 90
Telepathy, 13, 14, 23, 68, 87, 90, 108, 110, 120, 121
Transcendental Idealism, 25
Tremulation, 21, 23, 68
Trinity Doctrine, The, 8

V
Vedas, 25
Vital force, 21, 24, 66, 67, 70, 76

W
Witchcraft, 1

GPSR Compliance

The European Union's (EU) General Product Safety Regulation (GPSR) is a set of rules that requires consumer products to be safe and our obligations to ensure this.

If you have any concerns about our products, you can contact us on

ProductSafety@springernature.com

In case Publisher is established outside the EU, the EU authorized representative is:

Springer Nature Customer Service Center GmbH
Europaplatz 3
69115 Heidelberg, Germany

Lightning Source LLC
Chambersburg PA
CBHW072016290925
3332 7CB000 06B/678